NEUROSCIENCE RESEARCH PROGRESS

NEW DEVELOPMENTS IN SENSORY PROCESSING RESEARCH

Neuroscience Research Progress

Additional books in this series can be found on Nova's website under the Series tab.

Additional e-books in this series can be found on Nova's website under the e-book tab.

NEUROSCIENCE RESEARCH PROGRESS

NEW DEVELOPMENTS IN SENSORY PROCESSING RESEARCH

MARCEL VOISIN
AND
ROGER BRUNEL
EDITORS

New York

For permission to use material from this book please contact us:
Telephone 631-231-7269; Fax 631-231-8175
Web Site: http://www.novapublishers.com

NOTICE TO THE READER

Library of Congress Cataloging-in-Publication Data

ISBN: 978-1-62808-395-8

Library of Congress Control Number: 2013942942

Published by Nova Science Publishers, Inc. † New York

Contents

Preface

Sensory processing relates to the brains ability to organize and decipher different kinds of sensation entering the brain at the same time. In this book, the authors present current findings on new developments in sensory processing research. Topics discussed include descriptive sensory analysis of food; cellular principles of serotonergic modulation; sensory modalities processing in sensorimotor tasks depending on state and trait anxiety; and regulatory disorders of sensory processing.

Chapter I – Descriptive analysis is the most sophisticated of the methodologies available to the sensory professional when compared with discrimination and acceptance methods. All descriptive sensory tests involve the detection (discrimination) and the description of both the qualitative and quantitative sensory aspects of a product by trained panels. It is necessary that panellists demonstrate their ability to detect and describe the perceived sensory attributes of a sample. Majority of conventional descriptive profiling methods are based on fundamental issues, in particular the subjects selection process; the training phase including the development of descriptive vocabulary; and finally, the analysis of the data and information gathered. Descriptive analysis provides good quality data but, as a counterpart of this quality, it requires extensive training before the panel can be used as a reliable sensory instrument. These process steps make these methodologies are specially complicated and tedious, therefore require a high economic cost and time for the food companies. Recently, in response to industrial demands to develop faster and more effective methods of descriptive analysis new methods have emerged as alternatives to conventional profiling, these can be used either with trained or untrained panellists; this implies a reduction of the time spent and thus money invested. Choosing the right method for an evaluation depends on the aim and expected results of the evaluation, but, indeed, rapid methodologies are suitable to describe differences among products. The new methodologies provide product spaces that are generally comparable to those obtained with conventional descriptive analysis. Pros and Cons, origin, principles, statistical analysis and applications to food product are described in this review chapter.

Chapter II – By using small numbers of neurons, serotonergic systems in all animals modulate neural circuits from sensory inputs to motor outputs in an evolutionarily-conserved manner. While in mammals and, in general, in vertebrates the complexity of the nervous system makes it difficult to trace the effects of a neurotransmitter along whole neural circuits, the large and identifiable neurons of invertebrates allow cell by cell studies of the circuits that control stereotyped behaviors and of how serotonin modulates the performance of these circuits. Invertebrates such as decapod crustaceans, leeches and flies have allowed unveiling how serotonin regulates sensory inputs and their associated behaviors, including learning, social, feeding and circadian locomotory behaviors.

While serotonin released as transmitter from presynaptic endings acts in a milisecond time scale in hard-wired circuits, many of the effects of serotonin are slow and long lasting, characteristic of paracrine modulation and are presumably produced through the extrasynaptic release of serotonin from the soma, dendrites and axon varicosities. In particular, somatic release takes place over minutes following excitation and its effects may influence neuronal populations for long periods. The mechanisms of serotonin release from each neuronal compartment have been studied in leech neurons and later confirmed in mammalian neurons. Leech Retzius neurons receive inputs from interneurons and from sensory neurons. While interneurons drive low frequency firing that depends on the degree of electrical coupling between serotonergic neurons and evokes synaptic serotonin release, activation of pressure sensory neurons drive Retzius neurons to fire at higher frequencies that evoke somatic serotonin release. This chapter has as main goals: 1) to provide an overview about serotonergic neuromodulation of behavior from sensory inputs to motor outputs in crustaceans, leeches and flies, and 2) to review the cellular mechanisms that regulate serotonin release from synapses or from extrasynaptic sites, with particular emphasis on the neuronal cell body.

Chapter III – In the daily life activities, the ability to divide attention between different sensory modalities is essential since we have to cope with tasks that require multisensory integration. An individual has to select the relevant information, and sometimes shifts their attention from one sensory modality to another when one sensory modality is inadequate, absent, or salient. These sensory reweighting mechanisms are needed to maintain an efficient performance. Previous works have suggested that the highly complex attentional system that is needed in such situations could be influenced by affective states, such as anxiety. Indeed, it has been shown that an anxious state influences the efficient functioning of the attentional system. This can result in differences in the processing of sensory modalities. Moreover, recent works have highlighted that the allocation of attentional resources towards sensory modalities can be modulated by personality traits, such as trait-anxiety. This review focuses on the interdependence between anxiety state, anxiety trait and sensory modalities processing during tasks that require multisensory integration (*e.g.* sensorimotor tasks, balance control, posture...)

Chapter IV - Regulation Disorders of Sensory Processing (RDSP) is a clinical diagnosis in the *Diagnostic Classification of Mental Health and Development Disorders of Infancy and Early Childhood Revised edition: DC:0-3R* (Zero to Three, 2005). Data supporting RDSP as a distinct and valid disorder are scant. For this reason it was not included in the DSM diagnostic classification system. The aim of this chapter is to contribute to the validation of the sensory criteria of RDSP summarizing the findings from previous studies from the authors. In those studies it was analyzed in clinical samples: (1) the severity of sensory modulation deficits in children with RDSP; (2) the associations between sensory and behavioral symptoms; and (3) the specific role of sensory modulation deficits on the diagnosis of RDSP. Method: The samples were composed by toddlers (18-36 months-old) who were diagnosed by the clinician based on DC:0-3R diagnostic criteria. Toddlers were classified into two groups: those with RDSP and those with ‘‘other diagnoses in Axis I/II of the DC:0-3R’’(OD3R). Independently of the clinical process, parents completed the Infant Toddler Sensory Profile (as a checklist for sensory symptoms) and the Achenbach Behavior Checklist for ages 1 ½ –5 (CBCL 1 ½-5). The scores from the two groups were compared. Results: The results revealed that the RDSP group had more severe sensory modulation deficits (and more severe specific behavioral symptoms); stronger, although not significant, associations between most sensory and behavioral symptoms; and a significant sensory modulation deficits effect. Conclusion: The results confirmed the hypotheses, demonstrating the discriminative capacities of RDSP criteria versus OD3R, the independence of both sensory and behavioral criteria and the significant effect of sensory modulation deficits on RDSP diagnostic. The results also revealed convergent validity between the instruments and the diagnostic criteria. These findings suggest the importance of identifying RDSP in order to enhance the sensory capacities of children and to prevent clinical conditions at later stages.

In: New Developments in Sensory Processing … ISBN: 978-1-62808-395-8
Editors: M. Voisin and R. Brunel

Chapter I

Descriptive Sensory Analysis of Food: Conventional versus Innovative Methodologies

María José Jiménez[1*], Wenceslao Canet[2] and María Dolores Alvarez[2]
[1]USAS, Service Unit of Sensory Analysis
[2]Department of Characterization, Quality and Safety, Institute of Food Science, Technology and Nutrition (ICTAN-CSIC), Ciudad Universitaria, Madrid, Spain

Abstract

Descriptive analysis is the most sophisticated of the methodologies available to the sensory professional when compared with discrimination and acceptance methods. All descriptive sensory tests involve the detection (discrimination) and the description of both the qualitative and quantitative sensory aspects of a product by trained panels. It is necessary that panellists demonstrate their ability to detect and describe the perceived sensory attributes of a sample. Majority of conventional descriptive profiling methods are based on fundamental issues, in particular the subjects selection process; the training phase including the

[*] E-mail: mariajose.jimenez@ictan.csic.es.

development of descriptive vocabulary; and finally, the analysis of the data and information gathered. Descriptive analysis provides good quality data but, as a counterpart of this quality, it requires extensive training before the panel can be used as a reliable sensory instrument.These process steps make these methodologies are specially complicated and tedious, therefore require a high economic cost and time for the food companies. Recently, in response to industrial demands to develop faster and more effective methods of descriptive analysis new methods have emerged as alternatives to conventional profiling, these can be used either with trained or untrained panellists; this implies a reduction of the time spent and thus money invested. Choosing the right method for an evaluation depends on the aim and expected results of the evaluation, but, indeed, rapid methodologies are suitable to describe differences among products. The new methodologies provide product spaces that are generally comparable to those obtained with conventional descriptive analysis. Pros and Cons, origin, principles, statistical analysis and applications to food product are described in this review chapter.

Keywords: Check-all-that-apply; Descriptive methods; Flash profile; Pivot profile; Projective mapping; QDA®; Sensory characterisation; Sensory evaluation; Sensory profile; Sorting; Spectrum method

Introduction

According to Stone and Sidel [1], descriptive analysis is a sensory methodology that provides quantitative descriptions of products, based on the perceptions from a group of qualified subjects. Descriptive sensory analysis is the most powerful, sophisticated and extensively used of the methodologies available to the sensory professional (when compared with discrimination and acceptance methods) [1, 2].

All descriptive analysis methods involve the detection (discrimination) and the description of both the qualitative and quantitative sensory aspects of a product by trained panels. Panellists must be able to detect and describe the perceived sensory attributes of a sample. In addition, panellists must learn to differentiate and rate the quantitative or intensity aspects of a sample and to define to what degree each characteristic or qualitative note is present in that sample [3].

Traditionally, analytical tests (discriminative and descriptive) are performed with trained panels whereas affective tests are run with consumers [4].

Descriptive analysis is performed by a small number of panellists (from 8 to 20) who provide intensity ratings for a set of selected attributes. It involves a series of steps for the selection, training and maintenance of panel. Once the assessors of the panel are selected for each project or sample group it would: (i) development of vocabulary that describe the products, this is generally achieved by exposing assessors to many variations of the products and asking them to generate specific attributes that describe the similarities and differences between products; (ii) development of a consensus evaluation procedure about what the samples has to be tested; (iii) the panellists are trained in the evaluation and scaling of the selected attributes; and (iv) the assessors evaluate the products on the basis of each descriptive attribute, generally with the use of 10-15 cm unstructured line scales. The performance of the panel is monitored in terms of discrimination power, agreement between panellists and reproducibility during training to achieve the most accurate, reliable and consistent results as possible [5].

The data obtained from a descriptive analysis task would be in the form of intensity scores of all the attributes and the results enable one to relate specific ingredient or process variables to specific changes in some (or all) of the sensory attributes of a product [1]. The results provide the basis for mapping product similarities and differences, in which, by reducing the number of variables, it is possible to obtain one or more biplots representing the perceptual space of interest. The high specialization of descriptive panels allows obtaining very detailed, robust and consistent, reproducible results, stable in time and within a certain sensory space [6].

This methodology enables to measure the sensory reaction to the stimuli resulting from the consumption of a product providing a description of the qualitative and quantitative aspects of human perception [2]. From a product development viewpoint, descriptive information is essential in focusing efforts on those product variables that are identified as different among relative to a target, and also to establish causal relationships [1]. Food companies routinely use descriptive profile to define and quantify the sensory characteristics on which products differ because the information provided by descriptive profiles has numerous applications such as product development and improvement, quality control, advertising claim substantiation [4, 7], as well as understanding both consumer preferences [8] and their relationships with instrumental data [9].

Since the introduction of the Flavour Profile Method [10], several descriptive profiling techniques have been developed which can be found in the sensory evaluation literature. Some methods focus on the training of subjects on a sensory vocabulary, whereas other methods let subject more freely decide how to indicate differences between samples [11]. Among the methods more conventionally used are: Flavour Profile [10], Texture Profile [12], Quantitative Descriptive Analysis™ (QDA™, [13]), Spectrum Methods [3], Free Choice Profiling [14], Repertory Grid Method [15] and Time-Intensity Descriptive Analysis [16].

Definition of the sensory attributes of products has been an integral part of the food companies since long ago because information obtained from the description of the sensory characteristics of food and beverages enable companies to make more informed business decisions [4].

Conventional profiling methods, employing trained panellists, are commonly used to extract a descriptive profile of a food product. The methods are ideally suited for generating a detailed description of products [3]. However, they do not provide any indication of the importance of the attributes generated [17]. Thus, it may be difficult for product developers to know which attributes to focus most on.

Descriptive analysis provides good quality data but, as a counterpart of this quality, it requires extensive training before the panel can be used as a reliable sensory instrument. Because the vocabulary and associated training must be adapted to each product space, it can take form few weeks to several months to complete a study [5]. Therefore, this process can make these methodologies are specially complicated and tedious. Creating and maintaining a well-trained sensory panel can be quite expensive for both small and big food industries. For the first, this procedure implies a high economic cost, which could not support. The second have also to assume the consumption of an elevated time because these large companies have often a wide range of products that require various panels working in parallel. Furthermore, the training step can be relatively long as it should be detailed and extensive, particularly when industries require quick responses to market [7, 18].

For all issues considered above, is obvious the need for faster and more cost-effective methods which are more rapid and flexible, besides to give extra agility to sensory description and consuming less time. Moreover, to a traditional view on sensory evaluation, these methods are also the only tests where untrained individuals should be involved, whereas trained assessors are suitable for analytical sensory tests [4, 7].

In response to this demand, several methodologies have been offered as alternatives to conventional descriptive analysis. These alternative methods do not require a training phase and can be performed with assessors who have different levels of training. The innovative methods offer to the panellists the opportunity to select their own attributes identifying characteristics less technical and more similar to the vocabulary used by consumers. This matter never had been considered in the traditional approach.

In the last ten years, different techniques of sensory characterization had been developed which can be used with semi trained assessors (i.e. trained in sensory recognition and characterization but not in the specific category of products or in scaling) and even naïve consumers, with a great success, obtaining sensory maps very close to a classic descriptive analysis with highly trained panels [2].

Those innovative methodologies are, namely: Sorting [19, 20], Flash Profiling [21], Projective Mapping or Napping® [22, 23], Check-all-that-applies (CATA) questions [24], and other techniques less frequently used for sensory profiling (use of intensity scales with consumers, evaluation of open-ended questions, paired comparisons) or still in early development at this point in time as Polarized Sensory Positioning (PSP) [25], Preferred Attribute Elicitation Method [26], All-in-one test [27] and Ideal Profile Method [28].

These modern methods have been applied with consumers for gathering information about consumers' perception of the sensory characteristics of products that seem to be of great importance in sensory research. This issue is particularly important considering that food companies are increasingly basing their product development, positioning, advertising and communication strategies on consumer perception, in order to launch to the market products that are aligned as much as possible with consumers' perception [29].

In order to align products as much as possible with consumers' preferences and assure consumer' acceptance, industries need information about how consumers perceive the sensory characteristics of their products during product development [30].

Performing descriptive analysis with consumers, i.e., asking them to rate the intensity of different sensory attributes has been reported to be a good alternative to classical sensory profiles provided by trained assessors [31, 32]. According to these authors, sensory profiles obtained with consumers meet discrimination and reproducibility requirements, being an interesting alternative when companies face difficulties to use trained assessors' panels.

Sensory profiling of a product can guide product development teams on what to change to match the consumer's desired sensory profile, to get closer to a benchmark, to detect detailed differences created by a change of an ingredient, etc. [6]. Therefore, knowledge of "desired composition" allows for product optimisation and validated models between descriptive sensory and the relevant instrumental and/or preference measures are highly desirable and increasingly, are being utilised within the food industry [18].

There is plenty of literature extensively reviewing them in the last years. Murray et al. [18] have published a review on classical methods of descriptive analysis, mentioning the Flavour Profile Method [10], the Texture Profile Method [12], Quantitative Descriptive Analysis™ (QDA™, [13]), the Spectrum™ Method [3], Quantitative Flavour Profiling [33, 34] and what they called a "Generic Descriptive Analysis", which is generally a mixed approach taking ideas from some of these methods. On the other hand, Varela and Ares [2] have published the most recent review about novel methods for product characterization in which they cite: Sorting [19, 20], Flash Profiling [21], Projective Mapping or Napping® [22, 23], Check-all-that-applies (CATA) questions [24], and other techniques less frequently used as such Intensity Scales, Open-ended questions, Paired comparisons, Polarized sensory positioning (PSP, [25]) and Preferred attribute elicitation method [26].

This review chapter aims to discuss the implementation of both conventional and innovative methods of description sensory analysis. Therefore, origin, theory, applications, statistical analysis of the data, and advantages and disadvantages of each method are described in this chapter.

Conventional Methods

Flavour Profile Method

The Flavour Profile Method (FPM) was the first reported descriptive method, developed in the late 1940s at Arthur D. Little and Co. [10] to complement existing formal and informal sensory techniques for the expanding food industry [35].

FPM involves the analysis of a product's perceived aroma and flavour characteristics, their intensities, order of appearance, and aftertaste by a panel of four to six trained judges [3]. Although is one of the oldest techniques, FPM is still used frequently in industry particulary in flavour houses and the brewing industry [18].

Panellists, with the panel leader's help in providing and maintaining reference samples, develop and define the common terminology to be used by the entire panel. Panel leader carry out a personal interview to each panellist to determine her/his interest, availability and potential for working in a group.

Then panellists are selected on basis to her/his taste discrimination, taste intensity discrimination, and olfactory discrimination and description. Since FPM is a consensus technique, during both group discussion and training steps, panellists with the panel leader select the references standard; recorder the temporal order of the attributes; and define the common terminology and the scales to be used by the entire panel. For training, panellists are providing with a broad selection of reference samples representing the product range [3].

Once the panel has reached consensus, the panellists individually evaluate on sample at a time for both aroma and flavour, and they record the attributes (called "character notes"), their intensities, order of appearance and aftertaste. Finally, and after a general discussion, the panel arrives at consensus profile for each sample.

According to Murray et al. [18], an advantage of this methodologie is that the panel is highly trained and therefore sensitive to even small product differences. However, this advantage is also a limitation since the departure of one member of the panel can have a sever impact on the sensory results.Other disadvantage is that the technical language used by assessors may also be difficult to interpret by marketing personal in terms of relating the data to consumer preferences [18]. Other points of criticism of the FPM are, on the one hand, the seven-points scales used that it can limits the degree of discrimination among products showing small, but important, differences; and the other hand, that the panel's opinion may become dominated by that of a senior member or a dominant personality [3].

Texture Profile Method

Chronologically, the next descriptive method of importance is Texture Profile Method (TPM) developed by the Product Evaluation and Texture Technology groups at General Foods Corp. in the 1960s to define the textural parameters of foods [36].

Initially, Szczesniak [37] developed a texture classification system which proposed to bridge the gap between expert and consumer texture terminology, classifying perceived texture into three groups, "mechanical", "geometric", and "other" characteristics. Later, the method was expanded over the years by Civille and Szczesniak [38] and Civille and Liska [39] to include specific attribute descriptors for specific products including semisolid foods, beverages, and other products not foods. In all cases, the terminology is specific for each product type.

The technique aims to allow the description of texture from first-bite through complete mastication and also accounts for the temporal aspect of attributes [12].

According to Stone and Sidel [1], this method represented advancement in descriptive analysis from a structural point of view. They refer to the development of the descriptive terminology, the scales for recording intensities, and the word/product anchors for each scale category.

In developing the method, the objective was to eliminate problems of subject variability, allow direct comparison of results with known materials and provide a relationship with instrument measures [37].

Panellists are selected on the basis of their ability to discriminate known textural differences in the specific product application that the panel is to be trained for (solid foods, beverages, semisolids, etc.) [3]. In accordance with Lawless and Heymann [40], a minimum of 10 panellists are trained, with the number of training hours for a TPM panel being as many as 130 h over 6-7 month period.

Meilgaard et al. [3] stated that for training, the panellists are exposed to a wide range of products from the category under investigation to provide a wide frame of reference. The panellists define all terms and all procedures for evaluation reducing some of the variability encountered in most descriptive testing.

Each panellist independently evaluates samples. The original TPM used an expanded 13-point scale; however, TPM panels have recently been trained using category, line and magnitude estimation scales. As with the FPM, the panel verdict may be derived by statistical analysis of the data. For final reports, the data may be displayed in tabular or graphic form.

The main advantage is that the panellists are introduced to the underlying textural principles, this learning experience allow assessors avoid lengthy discussions about redundant terms and to select the most technically appropriate and descriptive terms for the evaluation of the products [3].

The extent of panel training in TPM may be perceived as a disadvantage; however, Otremba et al. [41] reported that this leads to greater consistency and accuracy by the TPM panel.

Quantitative Descriptive Analysis

The QDA method was developed in partial collaboration with the Department of Food Science at the University of California at Davis. In accordance with Murray et al. [18], quantitative descriptive analysis (QDA) was developed during the 1970s to correct some of the perceived problems associated with FPM [4, 13]. There were several distinct differences between FPM, TPM and QDA. Subjects for QDA methodology were recruited from sources removed from the project and were screened with dietary questionnaries and the products under test on the understanding that individuals who were frequent consumers of the product were more sensitive to product differences and thus more discriminating [42].

QDA method is based on the principle of a panellist's ability to verbalize perceptions of a product in a reliable manner; panellists are screened and trained in attribute recognition and scaling, they use a common and agreed sensory language, and products are scored on repeated trials to obain a complete, quantitative description [43].

Anyway, the language source in QDA is non-technical, everyday language, to avoid biasing response behaviour that may occur by providing a language, thus implying correct/non-correct answers.The panel leader is not an active participant in QDA to prevent bias and unstructured line scales are used to define/score the intensity of rated attributes [18]. Panellists require a certain level of practice. The panel is trained over a period of perhaps 10-15 h to understand the meaning of the attributes. QDA assumes that judges will use differentes parts of the scale to evaluate product attributes. The designs for descriptive analysis are based on repeated measures and the statistical analyses are generally conducted using Analysis of Variance. Often spider diagram is used to graphically represent the data. Precisely, in response to dissatisfaction among sensory analysts with the lack of statistical treatment of data obtained with the FPM or related methods, the Tragon Corp. developed the QDA method of descriptive analysis [4, 13]. This method heavily relies on statistical analysis to determine the appropriate terms, procedures, and panellist to be used for analysis of specific product [3].

A limitation of QDA is that it is difficult to compare results between panels, between laboratories, and from on time to another with this technique [18]. On the contrary, QDA training takes less time than other methods such as FPM or Spectrum method.

Panellists are selected from a large pool of candidates according to their ability to discriminate differences in sensory properties among samples of the specific product type for which they are to be trained. The training of QDA panels requires the use of product and ingredient references, as with the other descriptive method, to stimulate the generation of terminology. The panel leader acts as a facilitator. Attention is given to development of consistent terminology, but panellists are free to develop their own approach to scoring, using the 15-cm (6 in.) line scale that the method provides.

QDA represents a large step: the use of a graphical scale that reduces that part of the bias in scaling, resulting from the use of the numbers; the statistical treatment of the data, the separation of panellists durin evaluation; and the graphic approach to presentation of data have done much to change the way that sensory scientists and their clients view descriptive methodology.

The development of the method envolved from a number of considerations, including: responsive to all the sensory properties of a product; reliance on a limited number of subjects for each test; subjects qualified before participation; able to evaluate multiple products in individual booths; use a language development procuess free form leader influence; be quantitative and use a repeated trials design; and have a useful data analysis system [4].

Both QDA and TPM have been applied to study the effect of biopolymers and extra virgin olive oil addition in the sensory quality of mashed potatoes [44, 45]. On the other hand, also in potato puree, Jiménez et al. [46] has contrasted the applicability of the two innovative methods, flash profiling (FP) and projective mapping (PM), correlating the outcomes to a trained panel via QDA. Relative positioning of both samples and attributes in the three QDA, FP and PM sensory spaces was quite comparable. When clustering samples via Hierarchical Clustering Analysis (HCA), products were grouped in three clusters in the three methods. However, higher correlations between textural attributes evaluated by trained and semi-trained panels were found, indicating that although semi-trained assessors had not experience in potato puree samples texture description, they were able to evaluate them in a similar way as trained assessors did. Anyway, authors stated that difficulty with attributes coming from semi-trained and untrained assessors is that their real meaning and how they were measured is unknown, since only QDA panel knew the exact definition of all attributes.

The Spectrum Method

This method was designed by Gail Vance Civille in the 1970s [18]. Its principal characteristic is that the panellist scores the perceived intensities with reference to pre-learned, absolute intensity scales. The purpose is to make the resulting profiles universally understandable and usable at any laboratory outside the originating one [3].

The name Spectrum covers a procedure developed over the years in collaboration with a number of companies that were looking for a way to obtain reproducible and repeatable sensory descriptive analysis of their products [47, 48].

The principal tools are the reference lists together with the scaling procedures and methods of panel training. Spectrum is a "custom design" approach to panel development, selection, training, and maintenance [3]. For example, panellists may be trained to evaluate only one product or a variety of products, and these products may be described in terms of only appearance, aroma, flavour, texture, or sound characteristics, or in terms of all of theses attributes.

The method requires that panel that has been exposed to the underlying technical principles of each modality to be described. Panellists begin to develop their list of best descriptors by first evaluating a broad array of products that define the product category. Additional terms and references may be taken from the literature. The terms are then compiled and organised into a comprehensive yet not idiosyncratic list. This process includes using references to best represent the term so it is understood in a similar way by all panellists.

Different project objectives may require widely different intensity scales [3]. For this reason, 15-cm scales, a category with 30 points or more and magnitude estimatimation scale has been used. But, the most common application of the Spectrum scale is the use of 0-15 points for most foods and consumer products.

Murray et al. [18] stated that the use of references points along of the range of the scale, greatly reduces panel variability allowing for better correlations with other data, e.g. instrumental data. However, the Spectrum method requires an extensive training schedule and typical times required for each stage are 15-20 h for terminology development, 10-20 h for introduction to scaling and 15-40 h for final calibration.

Modified short-version Spectrum descriptive was developed by Muñoz et al. [49]. This version is based in the idea that certain applications of descriptive analysis require evaluation of a few detailed attributes without a full analysis of all the parameters of flavour, texture, and/or appearance. For example, the monitoring of product changes, necessary in quality control sensory work and in shelf-life studies, can provide the required information by logging a small number of selected sensory properties over time.

Free-Choice Profiling

Free-Choice Profiling (FCP) was developed by Williams and Arnol [14] at the Agricultural and Food Council (UK), complemented by the development of Generalized Procrustes Analysis (GPA) [50], a multivariate technique that adjusts for the use of different parts of the scale by different panellists and then manipulates the data to combine terms that appear to measure the same characteristics. These combined terms provide a single product profile.

The authors [14], claimed in discussing other methods that "complete agreement between assessors is often very difficult, if not impossible, to obtain". Therefore, they developed this methodology to assist the demands of marketing and product development teams who required information on target consumers' perceptions of products rather than the more technical descriptions of the products typically produced by trained sensory panels [18].

FCP allows the panellist to use as many terms as he or she needs to describe the sensory characteristics of a set of samples [51, 52]. The number of attributes generated is limited only by the perceptual and descriptive skills of the panellist [53]. The samples are all from the same category of products, and the panellist develops his or her own scoresheet.

The advantage of FCP is the avoidance of panel training, participants need only to be able to use a scale and be consumers of the product under evaluation [54]. A second advantage is that the panellists, not having been trained, can still be regarded as representing naïve consumers [3].

A disadvantage is that GPA reduces the information from studies to two or three dimensions; therefore, while FCP can reveal large differences between samples, it does not show the more discriminatory differences that would be revealed by conventional profiling (e.g. Cristovam et al. [55]).

However, by allowing panellists the freedom to select idiosyncratic attributes it may be possible to identify characteristics of products, which may not have been considered using a more traditional approach giving researchers new ways to differentiate products. At present, FCP is particulary useful for perceptual mapping of product spaces [40], and in situations where conventional profiling is not recommended, for example, Murray [56] used FCP to measure cross-cultural perceptions of snack-foods attributes in the English and Chinese languages.

Repertory Grid Method

The Repertoy Grid Method (RGM) is based on Kelly's clinical psychological Theory of Personal Constructs [57]. It is used in sensory consumer science with consumers in order to collect relevant and understandable information about what consumers perceive [15, 58, 59]. This method is particularly suited to vocabulary development and can therefore solve the difficulties in generating sufficient and suitable descriptors, a problem which usually arises when working with consumers [60].

In general, RGM is conducted in two sessions. The first one is dedicated to the attribute generation, in which the samples are presented randomly grouped in triads to asessors; then the panellists are asked to differenciate two samples from a third within each triad and explain in which way it is different from to other samples and also in what way these two samples are similar [15, 59-61]. The second session consists in rating the perceived intensity of the attributes generated in the first session for each sample. Thus, each assessor has his/her own list of attributes. The consumers enter their answer on individual score sheets.

The RGM seems to be efficient and easy to run, to such an extent that can be used with children between 8 and 11 years old [62].

The combined qualitative elicitation and quantitative ratings of the product characteristics can produce a product map with the aid of a multivariate analysis, such as Generalized Procrustes Analysis (GPA). The analysis of the data using GPA reduces the scale usage effects; delivers a consensus configuration in the use of attributes come from the usage of the same/similar attribute by different assessors; and allow comparing the proximity between the terms that are used by different assessors to describe products [6]. Therefore, one advantage of the RGM is that opens the door of the exploration of consumers' vocabulary with a more statistical validity.

As mentioned above, the RGM has already been performed with consumers but with the objective to understand consumers' perception of a product rather than comparing consumers' profiling with a trained panel's profiling and the aim of obtaining a product perceptual map [59, 63].

The RGM also presents the advantage of being also well suited as descriptive technique for a quantitative product mapping. RGM is recommended for obtaining a description when a quick profiling is needed or when the panel of trained assessors is not available or not trained in that particular product [6]. RGM has already been applied to the description of meat products [59] and bread [63].

Time-Intensity Descriptive Analysis

Perception of aroma, taste, and texture in foods is a dynamic, not a static, phenomenon. Time-intensity (TI) sensory evaluation is an extension of the classical scaling method providing temporal information about perceived sensations. By having judges continuously monitor their perceived sensations, from onset through extinction, one is able to quantify the continuous perceptual changes that occur in the specified attributes [16].

This analysis is different from the conventional descriptive analysis because it allows the monitoring of changes in the perception of an attribute over time. For certain products, the perception's intensity varies with time over a longer or shorter period, and an attribute's time-intensity curve may be a key aspect defining the product [64-68].

In the shortest term studies, completed within 1-3 min, the response can be continuously recorded.This response may be recorded using pencil and paper, a crolling chart recorder [64], or a computer system [69, 70] that is commercially available in several versions. The panellists should not see the evolving response curve being traced because this may result in bias from preconceived notions of its form.

Panellists may require several training sessions to develope and learn all of the protocols (for example, protocols for evaluation and for data entry) necessary for a well-controlled time-intesity study.

Time-intensity sensory evaluation has been normally used to evaluate single-attribute perceptions; however, a variation was presented by Duizier et al. [71], named dual-attribute time-intensity method (DATI).

This method is similar to the single-attribute test with the exception that two time-intensity scales, each representing one taste attribute, appear on the monitor. One taste is presented on the horizontal time-intensity scale, and the other taste is presented on the vertical time-intensity scale.

The results demonstrated that this variant allows collection of sensory data which more accurately reflects what is occurring in the mouth during consumption of a food. The success of the dual-attribute test was provided a tool for the study of many sensory interactions within foods.

Innovative Methods

Sorting

Classification, i.e. putting a group of things into categories according to an established criterion, is one of the most common operations in thinking. In the context of social sciences, the process by which a person classifies objects is called sorting [72].

Sorting has been extensively used as a systematic method for data collection in psychology, anthropology and sociology [73, 74]. The aim of the technique is to study how people classifies objects, which provides information of how people perceive the objects and what characteristics they attend to when making classification between a series of objects [75].

Lawless et al. [19] were the first to use this method with food products to get information about the sensory characteristics of food products in sensory and consumer science [76].

The sorting task aims to detect meaningful sensory characteristics within pairs of samples that explain similarities and dissimilarities within the investigated sample set. It consists of sorting products into groups according to their similarities. In general, sorting is performed in one session in which the samples are given all at once, then the assessors are asked to try the whole set of samples and to sort them into groups according to their similarities and differences, using their own personal criteria. The assessors place two samples in the same group if these samples are perceived as similar, whereas two samples that are markedly different should be placed in different groups. The asessors are free to make as many groups as they want and to put as many products as they want in each group. However, Moussaoui and Varela [6] recommended that the number of groups formed should be no less than 2 and no greater than 7, and besides that the assessors could not have only one sample in all groups.

In order to avoid trivial answers, assessors are usually told that they should sort the samples in at least two groups and that they could not have only one sample in all groups. In order to gather information about the sensory characteristics which are responsible for the similarities and differences between the samples, once the sorting has been completed, assessors are asked to provide descriptive words for each of the groups they formed [19, 29, 77-84]. This procedure is called 'labeled sorting' by Bécue-Bertaut and Lê [85].

When assessors are not trained, the description of each group generated may be difficult for them. Therefore, in order to facilite both this step and data analysis, Lelièvre et al. [84] provided to the assessors a list of predefined sensory characteristics and moreover these authors suggested to provide the assessor a predefined set of quantifiers to indicate the intensity of perceived attributes (e.g. "not", "a little", "medium" and "very").

The number of assessors used in sorting tasks depends on their training: when working with trained assessors the usual number ranges from 9 to 15 [78, 86, 87]. Despite the variability in the number of untrained assessors considered in sorting tasks, most studies work with 20-50 [6, 78, 88, 89].

Data analysis of sorting tasks aims at providing a spatial map that represents the similarities and differences between samples in terms of their sensory characteristics [19]. In this spatial representation the samples are represented by points on the map and the distance between a pair of samples is related to their degree of difference, which means that two samples which are represented close to each other are similar, whereas two samples which are represented far from each other correspond to dissimilar products.

The most common approach for analyzing data from sorting tasks is Multidimensional Scaling (MDS) [19]. When using this approach a similarity matrix is created by counting the number of times that each pair of samples is sorted within the same group. This similarity matrix is generally submitted to non-metric Multidimensional Scaling, but metric MDS can also be used because the sorting similarity provides a Euclidean metric [90]. Both non-metric and metric MDS are carried out in order to get a 2-dimensional representation of the samples.

The main disadvantage of MDS is that information about differences in the perception of the individual assessors is lost since the similarity matrix computes the similarity and differences between samples for the entire groups of assesors [19]. Therefore, it is not possible to determine if assessors sorted the samples similarly or if they had different perceptions and used different criteria to group them. To overcome this limitation, Abdi et al. [91], proposed the application of a different statistical technique, called DISTATIS. This technique takes into account the sample grouping provided by each assessor. DISTATIS first provides an optimal representation of the assessors based on their resemblance and also delivers a consensus sample representation. Finally, the words used by the asessors to describe the groups are projected by using barycentric properties.

Cadoret et al. [87] proposed the application of FAST for analyzing sorting data. FAST provides both samples and assesors optimal representations based on Multiple Correspondence Analysis (MCA) and Multiple Factor Analysis (MFA) respectively. In this technique all the assessors have the same importance when constructing the sample map.

The main advantage of DISTATIS and FAST is that they provide a representation of the assessors. Besides, the descriptors generated by the assessors could easily be projected into the sample space improving interpretation.

A first variation on sorting task, referred to as "directed sorting task" consists in providing information on either the number (e.g. the assessors may be asked to sort the products into two or three groups) and/or the nature (e.g. the assessors may be asked to sort-according to the colours red, rosé, or white-a set of wines that are presented in dark glasses) of the groups [86, 92]. This type is very useful for example to evaluate the effect of the region of origin of food products sensory characteristics [93].

Another variation first proposed by Rao and Katz [94] is called "hierarchical sorting task". Assessors are asked to form a given number of groups based on perceived similarities. Then, assessors are asked to successively combine the two groups that are most similar up to the time when a single group is formed (also called "ascendant hierarchical sorting" by Coxon [72]), or inversely to separate each group into finer groups up to the time when no further discrimination between samples is possible ("descendant hierarchical sorting") or both [95]. Ascendant hierarchical sorting has been applied to milk chocolate recently under the name of taxonomic free sorting by Courcoux et al. [96]. Descendant hierarchical sorting has been applied to olive oil by Santosa et al. [97] and to cards by Cadoret et al. [98]. Hierarchical sorting seems to give more precise information that sorting task as it gives rise to a more graduate measurement of the similarity between products. The position of the products on the sensory map could be more stable than that provided by sorting task [96]. But further studies are needed to validate this contention.

Sorting has several advantages which makes it an interesting methodology for sensory characterization. Firstly, it corresponds to natural and common mental activity, being an easy and enjoyable task for participants [72]. Besides, it does not require extensive training and produces little fatigue and boredom, which makes it appropriate for both trained assessors and consumers [99]. As suggested by Cartier et al. [78] another advantage of the methodology is that it does not require the use of scales or other quantitative systems.

The method also has the plus that it can be applied to a large sample set but it often needs to be completed by a verbalization task in order to describe the groups formed and to explain the dimensions of the resulting perceptual map [77]. However, it is important to take into account that all samples should be presented simultaneously in a single session, thus, when dealing with complex and fatiguing products, the number of products to be evaluated should be limited.

Despite a few differences, perceptual maps obtain with free sorting task (FST) are globally comparable with those obtained from descriptive analysis [29, 80] and seem to be reproducible [78, 84, 86, 89]. Nevertheless, the descriptions provided by untrained assessors could be difficult to interpret.

Sorting has been used on a large variety of food products including vanilla beans [100], cheese [19], drinking water [25, 89], fruit jellies [79, 83], beer [84, 101, 102], wine [85, 103], yogurt [80], spice aromas [104], cucumbers and tomatoes [105], breakfast cereals [78], olive oil [97], coffe [6] and orange-flavored powdered drinks [88].

Flash Profile

Flash Profile (FP) was suggested for sensory description by Dairou and Sieffermann [21] who defined it as an original combination of free-choice terms selection grouping with ranking method based on simultaneous presentation of the whole product set.

FP can be done in two sessions, or in one session with two steps. In practice, coded samples are presented all together. In a first step, panellists have to test them comparatively in order to generate all descriptors that they consider appropriate to discriminate between the samples. In a second step, they rank all samples from "least" to "most" according to each generated attribute, where ties are allowed. As in free choice profile, each assessor generates his/her own set of attributes individually, which should be sufficiently discriminant to allow a ranking of the samples; no indication is given regarding the number of attributes [6, 21, 106, 107]. The assessors are instructed to avoid hedonic terms.

When products are tested blind, and the samples permit (not easily recognizable by shape, colour, etc.), it is usual to include a repeated blind control within the sample set to examine individual assessor performance, subsequently checking in the final perceptual map that sample and blind control are close [108].

The simultaneous comparison of all the samples could allow better product discrimination. Furthermore, when the tested products belong to the same or to similar product categories, flash profling can be more discriminating than conventional profiling [106, 109].

Flash profile (FP) has been carried out with panels of 6 to 12 trained or semi-trained assessors [6, 21, 106, 110, 111] and with consumer panels of 20 to 40 participants [6, 107, 112].

For the data anlysis, individual matrices for each assessor (samples x attributes) are built in order to enter product rankings. A Generalised Procrustes Analysis (GPA) [50] is then performed on the matrices in order to obtain the sample and attribute configurations. It is important to point out that GPA reduces the scale usage effects, delivers a consensus configuration and also allows comparing the proximity between the terms that are used by different assessors to describe products [6], so is very suitable to this methodology.

A variant of FP called Ranking Descriptive Analysis (RDA) has been recently proposed by Bragato Richter et al. [113]. This variant is based in previous work by Rodrigue et al. [114]. In RDA, assessors first develop a consensual vocabulary. This procedure involves four steps: (i) attribute development with selection of the most frequent terms; (ii) definition of the selected terms; (iii) development of a consensus evaluation procedure; and (iv) assessors are asked to rank the samples for all the selected terms. RDA can be seen as a compromise between descriptive analysis and flash profile. Although not as fast as flash profile, it preserves the idea of using a consensual vocabulary and this facilitates data interpretation.

Between the advantages of FP is: to be a rapid sensory mapping tool, easy to comprehend for panellists, suitable for applications when a quick response is needed but also as initial screening tool for a new product set or category and to study a specific market [21, 106, 111].

A limitation of FP, being a comparative method, is that the number of samples that can be assessed is limited, and it would depend on the product category. Another drawback of FP is that each assessor has his/her own attributes list, so the semantic interpretation can be complex [21, 112]. However, it has been proved that even if this method produces a large amount of varied attributes, the cores attributes for the description of the samples set are well covered when using FP even with consumers speaking different languages, what makes the method especially suited for cross country comparisons [6].

Flash Profile has been used to description of food products such as fruit dairy products [106], red fruit jams [21], potato puree [46], commercial apple and pear purees [111], lemon ice teas [112], fish nuggets [110], bread [107], hot beverages [6], chewing gum [115] and Cuajada cheese [116].

Projective Mapping and Napping®

Projective Mapping (PM), and its variant Napping®, are profiling methods that were developed [22] in order to collect a Euclidean configuration for each assessor in a single sensory session. There were originally derived from psychology and previously used in qualitative market research, to obtain associations between products [22, 23]. PM was proposed to use it with consumers, and linked the results to trained panel data to explain product description [23].

The samples are simultaneously presented to panellists, to be positioned by each assessor on a bi-dimensional space as a "tablecloth" ("nappe" in French, which gave the name to Napping®), or more often in an A4 or A3 blank paper. Samples are arranged by the participant according to the differences and similarities between them in such a way that the smaller the distance separating two samples, the more similar they are [17]. The positioning criteria and their importance are chosen on an individual basis by each assessor, which makes PM a flexible and spontaneous procedure [6]. The simultaneous processing of all these configurations provides a graphical display of the product in which two products are near if they were perceived similar by the whole panel of subjects [22].

After positioning the samples, the participants are sometimes asked to describe the individual samples or groups by directly writing on the sheet, as a means to describe the differences or sensory attributes, the reasons behind their placement, method generally known as "Ultra Flash Profiling" [17, 117].

When the exercise is blind, and the samples allow (not easily recognizable by shape, colour, etc.), it is common practice to add a repeated blind control sample in the set, to check assessor performance, subsequently checking in the final perceptual map that sample and blind control are close [108].

Via multivariate statistical analysis of the data, all the individual maps are collated in a consensus configuration that, together with the comments, woul determine the sensory profile of the sample set, in terms of distances and descriptions, as perceived by the group of assessors.

When working with trained panels, PM has been performed with 9 to 15 assessors, as in traditional descriptive analysis [17, 23], whereas when untrained assessors or consumers complete this type of task the number has been increased to 15-50 [110, 118, 119].

Data are collected manually as the x- and y-coordinates to the location of the samples for each assessor. Together with samples coordinates, the comments given for each of the samples are listed together with their frequencies of mention, resulting in three tables with the samples in rows and the coordinates (x, y) and the mentioned attributes in columns. The projective maps from each panellist obtained by the PM are analyzed by Multiple Factor Analysis (MFA) [120], in which the simultaneous processing of all these configurations provides a graphical display of the products (biplots) in which two products are near if they were perceived similar by the whole panel of subjects [22].

MFA is an enriched PCA which analyzes several tables of variables differing in number and nature, with the requirement that within a table, that the variables must be of the same nature (quantitative or qualitative). MFA integrates different tables of variables describing the same observations [6, 110]. The difference with PCA is that MFA takes into account individual differences rather than averaging the data [119]. When attributes generates by ultra flash profiling are incorporated, the qualitative data is analyzed as another data table that is accounted for as supplementary variables [17]. Moussaoui and Varela [6], Pagés [22] and Perrin et al. [17] present schematic exemplifying views of how to collect and treat the Napping® data.

Napping® is a flexible method therefore it is a technique that still today is developing to adapt it to different objectives and/or depending on the complexity of the products. Pagès et al. [121] proposed the Sorted Napping® procedure which combines napping with a categorization task consisting in that after the Napping® exercise panellists are asked to identify explicit groups by circling products on the sheet. Also, in its initial paper, Pagès [117] suggested as a good idea to perform napping by modality (appearance, odour, flavour, texture), this variation is known as Partial Napping which was later on implemented by Pfeiffer and Gilbert [122] and presented in Sensormetrics2008. Finally, Ferrage et al. [108] suggested in their communication in Eurosense2010 the possibility of performing Structured Napping, where pre-named axis would be given to the panellist, signaling the 2 main directions of the sensory flavour space (x and y axes of the "nappe"), this variant has not been further investigated so far.

Napping® is based on the global perception of the sample set differences, and it has been described as a natural, intuitive, holistic way for consumers to describe products, closer to what happens in front of the supermarket shelf [88, 123].

Napping has the advantage to be quite easy and understandable for consumers to perform. Risvik et al. [124] even suggested the possibility to use it with children, for the possibility of turning it into a game. However, Ares et al. [88], report that other novel techniques could be easier to understand than Napping® to naïve cosumers.

The highest disadvantage of napping is the number of products that can be tested at the same time that would much depend on the sensory characteristics of the product, but generally with a maximum of 12 [22]. Other limitation is the reproducibility, validity or robustness of the methods, like most novel techniques, has not been studied in detail so far [2].

This methodology has been applied to diverse food products like chocolate [23, 125], commercial dried soup samples [124], snack bars [126, 127], ewe milk cheeses [128], citrus juices [119], wines [129], hot beverages [6], fish nugget [110], apple and cheese [130], milk desserts [118] and powdered drinks [88] and potato puree [46].

Particularly, Moussaoui and Varela [6] explored the performance of both FP and PM methods and their linkage to a QDA among target consumers of hot beverages. They concluded that these techniques could be seen as complementary to the information given by a trained panel, being very powerful tools to obtain direct feedback from consumers. The same authors stated that next steps would be to study "how far can these methods go", i.e. would be they applicable to most product categories? Following, Albert et al. [110] used QDA, FP and PM to study their suitability on hot fish nuggets. Maps of the sensory spaces obtained by the three methods were well correlated showing that FP and PM could be used as quick alternatives to QDA in these products. Recently, Jiménez et al. [46] also compared trained, semi-trained and untrained sensory profiles for seven potato puree samples by using quantitative descriptive analysis (QDA), flash profiling (FP) and projective mapping (PM), respectively. Results showed that both FP and PM methodologies were able to identify differences in sensory characteristics of potato puree samples, providing very similar sensory maps. Nevertheless, higher correlations between sensory attributes evaluated by semi-trained and trained assessors were found, indicating that preferably FP should be used as a rapid choice to QDA in hot served potato purees.

Check-All-That-Apply Questions

Check-all-that-apply questions (CATA) was originated from the work of Coomb [131] and then was proposed in marketing research as simple method to gather information about consumers' perception of the sensory characteristics of food products and to help optimising food products [24, 132].

CATA questions are multiple choice questions which are commonly used in marketing research in order to reduce response burden [133]. These questions consist of a list of attributes (words or phrases) from which assessors should select those they consider appropriate to describe a product [5].

In this methodology, samples are presented to assessors in monadic sequence, following a balanced rotation order. Panellists are asked to try the samples and to answer a CATA question by selection all the terms that they consider that apply to answer this question in order to describe each of the samples. There are usually no restrictions on the number of attributes that assessors can select, therefore they can check as many attributes as they wish and they can take as much time as needed. The list of words or phrases in the CATA question could be exclusively related to the sensory characteristics of the sample or also include terms related to non-sensory characteristics, such as usage occasions, product positioning, emotions or concept fit [134].

The terms of the list given to assessors could correspond to the descriptors used by trained assessors panels to characterize the products or could be selected considering results from previous focus groups or quantitative consumer studies [2]. On the other hand, the terms can also be derived form the results of open-ended question [135]. Open-ended questions are usually used in consumers test in addition to a hedonic scale to obtain a rough product description. After evaluating their liking of the products, consumers are asked to provide a few terms to either describe the products or to indicate what they like or dislike about these products [135-137].

As this methodology is mainly used with consumers, the number of assessors necessary to perform a sensory characterization using CATA questions rages from 50 to 100 [131, 134, 138].

A first step in the analysis of the data from CATA questions is performed using Cochran's Q test [139] in order to determine if assessors detect significant differences between samples for each of the terms. This analysis is a nonparametric statistical test used in the analysis of two-way randomized block designs to check whether k treatments have identical effects, when the response variable is binary [2]. For each term of the CATA question a data matrix is created containing samples in columns and assessors in rows. In this matrix each cell indicates if the term have been checked or not (1/0 respectively) by each assessor to describe each sample.

Multiple Correspondence Analysis (MCA) [88, 140] can then be performed on a matrix containing the number of assessors who checked each term from the CATA question. This analysis is carried out to obtain a sensory map of the samples, which enables to determine the similarities and differences between the samples, as well as the sensory attributes that characterize their sensory attributes [135, 138].

Recently, Popper et al. [141] have also suggested using a variant of Correspondence Analysis called multi-block Hellinger analysis that relies on the Hellinger/Bhattacharyya distance [142] rather than the chi-square distance of Correspondence Analysis and that can also integrate individual differences.

CATA questions have been used for the sensory characterization of several food products for example: salty snacks [24, 141], strawberry cultivars [143], vanilla ice cream [134], chocolate milk dessert [135, 138], texture perception of milk desserts [144] and citrus-flavoured sodas [145].

A variant of CATA, called 'pick-*K* attributes', is also used in both marketing and sensory domain. In the food sensory domain, this method had been mostly applied to describe aroma of complex products such as wines [146, 147]. Assessors receive also a list of attributes and they choose the *K* attributes that are dominant or that describe best a product. The main difference between this variant and CATA is that when *K* is small, the pick-*K* attributes method highlights the main sensory characteristics of the products, whereas the CATA method provides a more complete description.

The main advantage of CATA questions is that allows multiple options to be selected, instead of limiting respondents to select only one answer or forcing consumers to focus their attention and evaluate specific attributes [148].

Some studies have compared the sensory maps generated by CATA questions with those provided by classic descriptive analysis (QDA) with a trained assessor panel, reporting similar results [134, 138, 144]. According with these authors, CATA questions have been reported to be a quick, simple and easy method to gather information about consumer perception of the sensory characteristics of food products. Moreover, Adams et al. [24] have reported that CATA questions have a smaller influence on liking scores than just-about-right or intensity questions.

However, it is important to take into account that despite the fact that frequency of mention of the terms from CATA questions not provides quantitative information since consumers only evaluate if a term is appropriate or not to describe the product [2].

Sudman and Bradburn [149] indicated that when CATA was used in marketing research, respondents may not select a term in a check-all question due to three main reasons: because they consider that the term does not apply, because they are neutral or undecided about it, or because they did not pay attention to it. Therefore, if consumers do not select a term in a CATA question it cannot be concluded that they consider that it does not apply to the product.

Moreover, it has been reported that respondents tending to select the terms that appear near the top of the list rather than those that appear near the bottom of the list [150]. For this reason, rotation of terms within a CATA question is recommended for getting valid results. However, further reseach is needed to evaluate the influence of the number and type of terms, as well as the design of the questionnaire and results from sensory characterization provided by this methodology [2].

Another limitation of CATA questions is that they provide qualitative data and therefore might have smaller discriminative capacity than ranking tasks or intensity scales [134].

Campo et al. [151], have recommended that this approach could be a practical alternative to conventional descriptive analysis for characterizing products with complex aroma. Also Nicod et al. [152] recommended the application of CATA questions with trained assessors for the evaluation of complex products. In particular, they suggested its use for attributes present in the sample in low concentrations. These attributes cannot be easily discriminated via scaling since their perception is more related to presence/absence than to different intensities.

This approach is not recommended when evaluating products with slight differences in their sensory characteristics because the same terms will be selected for all the evaluated samples and therefore would not be able to detect significant differences between samples [2].

Other Methodologies

a) Intensity Scales

This methodology is performed with untrained panellists, i.e, consumers. In this approach consumers are asked to rate the intensity of a fixed set of sensory attributes. Commonly, for rating these attributes is used structured scales anchored with "nil" and "very" [30, 144]. The main difference with the traditional approach is that descriptors are provided to consumers by the researcher. To select these descriptors, either focus group [30] or CATA questions [144] methods can be performed. Moreover, other dissimilarity with conventional descriptive methods is that no training in attribute recognition or quantification is performed.

Generally, for the data analysis obtained with this methodology, mean, standard deviation, minimum and maximum scores are calculated for each of the descriptors evaluated using intensity scales. Then, ANOVA is carried out on consumers' intensity scores for the evaluated attributes considering samples and consumers as sources of variation, and using a post hoc test (Tukey's test, for example) with a significance level appropriated. Later, Bruzzone et al. [144] performed a principal components analysis (PCA) on the mean values of the evaluated attributes to generate a sensory map of the samples, whereas that Parente et al. [30] performed a multiple factor analysis (MFA) on the means of consumer's data.

Despite this approach has been traditionally not recommended [1, 7], recent studies have reported that results from sensory characterization performed by 50-100 consumers with intensity scales are similar to those provided by trained assessors panels [31, 32].

However, Ares et al. [153] compared global and individual performance of a consumer and trained assessor panel concluding that consumer agreement was low and the majority of consumers were not able to give scores that significantly discriminated among samples.

Other authors [30, 144] also have indicated the disadvantage of the large variability in consumers' rating because of the lack of training. This should be taken into account despite the fact that this variability is compensated by the large sample size that should be used in this type of the studies. For these reasons, the evaluation of sensory attributes using intensity scales by consumers should be gathered as complementary to quantitative descriptive analysis performed with trained asssesors [30]. Nevertheless, intensity sales could be a valuable alternative to gather information about the sensory characteristics of food products, particulary interesting for companies which do not have trained panels or which do not have enough time or resources to train the panel for a specific application [30, 144].

Therefore, some authors have stated that the training period could be omitted and have performed sensory characterization with the use of consumers [31, 32].

b) Open-Ended Questions

This methodology is usually used in studies with consumers in order to describe the sensory characteristics of the set of products. Its aim is understand the main characteristics that determine consumer perception of the products and especially what they like or dislike about these products.

Different options have been studied that enables consumers to provide a description about the sensory characteristics of food products. ten Kleij and Musters [137] allowed consumers to voluntarily write down remarks after their overall liking evaluations. In the study performed by Ares et al. [135], the consumers provided up to words to describe the samples after their overall liking evaluation. More recently Symoneaux et al. [136] gave consumers the option to freely state what they liked and/or disliked about the evaluated products.

The number of assessors necessary to perform this methodology ranges from 50 to 100 [135-137].

A disadvantage of open-ended questions is derived of the large amount of textual data obtained which are often difficult to analyse. This analysis is labor-intensive and time-consuming. A first step consists of deleting auxiliary terms and other irrelevant words. Then, words with similar meaning are grouped into the same category. This classification is usually performed consensually by three researchers [154]. Categories mentioned by more than 5% or 10% of the consumers are selected and frequency of mention of each category is determined by counting the number of participants that used each category to describe each product. A frequency table is constructed and analysed using Chi-squeare test and correspondence analysis. Global Chi-square test could be used to determine significant differences in the description of the evaluated samples by studying the independence between rows and columns [135]. Moreover, Chi-squares can be used to identify significant differences between samples for each of the sensory charcteristics used by consumers to describe the evaluated products [136]. Finally, correspondence analysis can be used to get a 2-dimensional representation of the samples and the attributes [135, 137].

The sample maps gathered from open-ended questions have been reported to be similar to those obtained from quantitative descriptive analysis (QDA) with trained assessor panels [135-137].

Open-ended questions are not subjected to limitations from the researchers and therefore with this methodology can be obtained rich information that could underscore and complement quantitative data from trained assessors [137]. However, the information provided by this methodology is not as precise as the information provided by descriptive analysis or other methodologies, particularly in those cases in which differences between the products are small [2]. The consumers' responses to open-ended questions could be interesting when designing communication strategies [135, 136].

A variation of the open-ended questions is a method called Pivot© profile which was proposed by Thuillier [155]. Pivot© profile was developed in the field of wine description to characterize the sensory properties of champagne with experts. The strategy implemented in this method to simplify data analysis and to reduce inter-individual variability is to collect free descriptions of the differences between two products: a sample product and a single reference product (the so called "pivot").

In Pivot© profile, assessors are first provided with the reference product, and then, the products to be evaluated are presented one at a time. Assessors evaluate all products and the reference product and they write down each attribute than the product has in smaller or larger amount than the reference product (e.g. less sweet, more astringent). Assessors only use descriptive words without any sentence, and the negative form is not allowed (e.g. flat for non-effervescent).

This variation of open-ended questions seems to be a very promising method for complex products such as wines where a tradition of free description is relatively strong among experts. On the other hand, the main difficulty remains to choice of the reference product, and, here also, further work is needed to explore this issue.

This methodology has been used in a limited number of food products: mayonnaise [137], milk desserts [135, 156] and apples [136].

c) Preferred Attribute Elicitation Method

Preferred Attribute Elicitation (PAE) is a novel approach proposed by Grygorczyk et al. [26]. In this method, a group of consumers is asked to agree on a set of attributes to describe a sample of products and to rank the attributes according to how important they believe the attributes are in influencing their liking of the products.

The PAE method was previously described by Nash in a conference poster (*Understanding Consumer Perceptions, Atttitudes and Choices for Soymilk Beverages*. Conference poster presented at The 5th Pangborn Sensory Science Symposium in Boston, MA, 2003) [26].

The study was performed in samples of vanilla yogurts that were selected from a pool of 19 commercial yogurts. Panellists evaluated texture liking, flavour liking and overall liking of each yogurt on a 9-point hedonic scale.

Then, they wrote down in words what attributes they liked or disliked about the yogurt to encourage panellists to start thinking about yogurt attributes. Afterwards, through round-table discussions moderated by the researcher, panellists were asked to name attributes which differed among yogurt products and to group the generated attributes into groups in any way they consider appropriate. Next, 7-point scales were generated for the attribute groups and panellists assigned anchor descriptor terms for intensity for each attribute. Subsequently, panellists were then asked to rank the attributes governing each scale according to importance in driving liking. After a short break, panellists evaluated each attribute for each yogurt on a 7-point scale with the anchor terms they had selected.

The descriptive data acquired were collected from the different PAE groups and combined and normalized using Generalized Procrustes Analysis [26]. This study was performed with 42 consumers and the results were compared with those obtained from a conventional profiling with 10 trained assessors, and theses results were similar.

According to the authors, the advantage of this method is that the method would be useful for product develops in order to search a handful of key attributes to focus their efforts. On the other hand, PAE does not provide a detailed description of products, thus the method has more applicability in prior steps of the product development process.

d) Polarized Sensory Positioning

This methodology was developed by Teillet et al. [25] in order to rapidly reveal the sensory characteristics of water, without the need for a large number of samples. This method is simply based on the comparison of samples to water standards. The objective of Teillet et al. [25] in their research was proposed a new sensory methodology well suited to the determination of the taste of water.

The samples used in the test were chosen from the most popular brands available on the French marked. Despite the fact that these authors applied polarized sensory positioning (PSP) with 15 trained assessors, this methodology could be applied with semi-trained assessors or naïve consumers because this method could be used without special training of panellists.

The methodology simply involved measuring the sensory similarities of a water sample with the prototypes ("poles"), i. e, the panellists were asked to provide a score relative to each pole on a continuous scale ranging form "exactly the same taste" to "totally different taste".

For the data analysis, there was two different ways to process; the first consisted in used Multidimensional Scaling unfolding techniques [157] considering scores given on the scale as distances between products in order to get a two-dimensional representation of the samples. In the second approach the poles are considered as "global" descriptors, then data are analyzed by calculating average scores, and sample representation is obtained by Principal Component Analysis [25].

Polarized sensory positioning is an easy, quick methodology and provides promising results, but there still remain questions to improve the use of the technique such as: (i) from point of view semantic, the anchors of the scales used have a problem since are inherent to each individual; and (ii) the choice of the number of poles to use and their sensory characteristics are important factors which could be considered.

e) Paired Comparison

Paired comparisons are one of the most common methodologies in sensory science [7]. The aim of this method is to determine if two samples are identical or different to a specific attribute. Generally, two samples are presented to panellists in order to compare them, afterwards the panellists are asked to determine how a particular sensory characteristic differs between two samples (for example, "determine which sample is the sweetest").

This methodology has been used in non-food products. Poirson et al. [158] proposed the application of this test for gathering information about consumers' perceptual space of diesel motor sounds. These authors asked panellists to perform a series of paired comparison tasks according to a list of attributes and to assess their difference by using a 7-point scale. The authors also developed rating tasks with naives' consumers and conventional sensory profile with experts.

Data of paired comparisons were analyzed to determine the discriminative power of the attributes and the consensus between panellists by using least squares logarithmic regression (LSLR).

Poirson et al. [158] reported that paired comparisons provided better agreement between consumers and were more discriminating than ratings tasks. The results of these two tests with the naives were then compared with the comventional sensory profile of the experts using Generalized Procrustes Analysis. GPA revealed good consensus between results from paired comparison task and those from descriptive analysis with a trained assessor panel.

f) All-In-One Test

Methods to collect sensory profiles relying directly on consumers' perceptions are increasingly employed. One of these methods has been developed by Giacalone et al. [27], and it applied in a study of beers.

Initial results with "All-In-One Test" (AI1) test show that it is an efficient and versatile approach for exploratory product testing with consumers. This exploratory AI1 provided interpretable results concerning consumer perception (sensory/hedonic) of the beers, and revealed relations with consumers' background information.

The aim of this study is to further increase the potentials of combining different domains of consumer product perceptions simultaneously in one test method. This test combines several existing techniques for collection of consumer data into a unique experimental approach, and aims at simultaneously collect information about consumer background, appropriateness for specific sensory properties, hedonic ratings of food products and a sensory profile by Check-All-That-Apply (CATA) technique using consumers as sole data source [27].

Six commercially available beers were used in the study, which were chosen to represent different beer styles offered in the Danish beer marker. The test included two consecutive steps: a conceptual part by means of a questionnaire, and a sensory part. First, consumers provided information (in a questionnaire) about: (i) specific demographic (gender, age, education level…) and psychografic (eg. interest in beer) data, and (ii) appropriateness ratings for 27 sensory descriptors for an "ideal" beer (eg. bitter, nuts, and fruit). Secondly, in a sensory part, consumers were presented with six beer samples (in a monadic sequence), for which they had to score perceived liking on a 7-point hedonic scale, and provide a sensory profile of the beer by answering to a series of CATA questions were descriptors were exactly the same as the one used in the preliminary questionnaire.

For the data analysis were carried out both univariate and multivariate analyses. The first consisted in analyses of variance performed on the appropriateness rating and on the hedonic rating. Afterwards, for the multivariate analysis, the collected data were organized into three data matrices: the X matrix contained the sensory data gathered from the consumers; the matrix Y contained consumer liking data; and the matix Z contained the consumer background information collected by the questionnaire, i.e. demographic, psychographic and appropriateness data. The resulting three blocks data structure was analyzed by L-shaped Partial Least Square Regression (L-PLSR) [159].

According with the authors [27], the AI1 test presents some potential issues that need further exploration. The first one is the data produced will generally have a lot of noise and variability due to use of consumers for descriptive analysis. Secondly, careful consideration should be given to attribute selection when preparing CATA questionaries. Thirdly, CATA may present some limitations compared to rating attributes since the use of nominal data requires a more prudent interpretation. The last methodological issue concerns the use of AI1 test as a tool for investigating consumer preference suggesting that asking hedonic ratings first (as authors have done in AI1 test) coul limit the effects on liking [160].

For all that is mentioned above, AI1 test can be considered as a fast and easily applicable tool for sensory and consumer scientists to support explorative stages in consumer-oriented food product development. Although more systematic studies are necessary, the authors see AI1 as an opportunity for growth of the sensory science field to better support product developers.

g) Ideal Profile Method

The Ideal Profile Method (IPM) is a technique which aims at acquiring sensory data from consumers [161]. In this methodology, the samples are presented in a sequential monadic order taking care of order and carry over effects [162] to consumers. They are asked to describe the samples according to a set of given sensory attributes for both perceived and ideal intensities. In addition, the consumers are also asked to provide hedonic ratings of the products. Therefore, the most important profit of this method is the large data provided, that is, are gathered three types of information from each consumer: the sensory profiles of the products (i.e. how consumers perceive the products), the vector of hedonic scores (i.e. how much consumers like the products) as well as the ideal profiles (i.e. what are the consumers' expectations) [163].

In the *ideal* situation, when consumers describe their ideal profiles, they describe *fictive* products, therefore, the ideal products provided by consumers should have a higher liking potential than the products themselves [161].

In order to better understand which sensory attributes influences positively or negatively the liking of the consumers, an analyisis joining sensory description and hedonic judgments can be performed. This is the principle of the *preference mapping* techniques [8] which became important in the process of product development as it guides for products' improvement.

Therefore, analyses such as Multiple Factor Analysis (MFA, [164]), Partial Least Square (PLS) regression [165] have been proposed. However,

Worch et al. [28] proposed a variation of preference mapping, the *Ideal Mapping*, when sensory data is collected according to the Ideal Profile Method. This technique takes into consideration the sensory profiles of the products and the ideal profiles directly provided by consumers for the construction of the map. The Ideal Mapping (IdMap) technique is similar to the external preference mapping technique, however the nature of data is different in both: for external preference mapping, the sensory profiles are combined to the hedonic scores while in the IdMap, the sensory profiles are associated to ideal ratings. For this reason, the IdMap aims at defining (estimated or directly) the profile of an ideal product that could be used to guide product developers for the improvement of their products. Thus, the IdMap is rather a complement than a substitute for the preference mapping [28]. It is therefore recommended by these authors to combine both techniques, so that IdMap can enrich preference mapping.

The Ideal Profile Method can help manufacturers to improve existing products [166]. However, the ideal descriptions obtained are delicated as are come directly form consumers; therefore, to avoid any possible misinterpretation of this data which could lead to an incorrect reformulation of the tested products, it is important to check the consistency the ideal data before use. The process checking for the consistency of the ideal profiles was proposed by Worch et al. [161].

Because this method has recently been developed, it has mostly been applied in luxurious perfumes [167].

Conclusion and Outlook for the Future

Descriptive analysis is undoubtedly one of the most valuable tools in the field of sensory analysis and is extensively used by many sensory professional. In the next millennium, it is expected that descriptive analysis will be used increasingly for a wide range of end used than ever before [168]. Considering this, it is vital that investment continues in the development of descriptive analysis that challenges traditional ideas in order to ensure optimal potential is gained from this method in the future [18].

Despite the fact that the most of the novel methodologies proposed in this chapter seem promising, further work is clearly needed to better characterize them. In particular, unified tools are still needed for evaluate the reliability of these alternative methods. Some authors use repetitions (e.g. [78, 84, 89]), some other authors use duplicate samples (e.g. [6, 86]), other authors use R_V coefficients to compare maps obtained by several panels [6, 21, 83, 84, 86, 106, 110, 125], whereas other authors use bootstrapping to draw confidence ellipses around the products [92, 97] on the sensory map. Blancher et al. [169] propose to use resampling techniques and to draw large numbers of samples of different sizes from the original set and to compute the average R_V coefficient (which they called R_{Vb}). These authors suggest considering these techniques as reliable experiments for which this average R_{Vb} is at least equal to 0.95.

Innovative methodologies consist of valid, reliable, simple and quick alternatives for sensory characterization of food products. Moreover, the hypothesis that consumers are able to accurately describe products is more and more accepted within the sensory science community and diverse product profiling methods are never before in the food industry [2]. On the other hand, the results obtained from the novel methodologies are similar to those gathered by conventional methods using trained assessors, even the sensory spaces are more or less analogous [6, 21, 23, 25, 46, 78, 86, 106, 110, 124, 134-136, 138, 140, 144]. However, it is important to emphasize that the recent methods of description sensory have not considered a replacement for classical analysis; the reason for this could be that the conventional procedures are always more accurate since the assessors are extensively trained in the identification and quantification of sensory attributes according with the specific study. Therefore, the new methods have emerged more as complementary rather than substitutes.

Both conventional and innovative methods of description sensory are appropriate and used for determined applications. Classical descriptive analyses are more appropriate when the objectives of the sensory characterization are: (i) to identify small differences between products or to detect differences in the intensity of specific sensory attributes (it occur during the development of new products, concretely in the optimization step, for example when it tested the change of an ingredient in a food; (ii) to compare samples in different moments in time; and (iii) when a very detailed sensory description is required.

On the other hand, when non-detailed information about the sensory characteristics of food products is required, novel methodologies offer a good and quick alternative [2]. They are useful: (i) to conduct preliminary studies for getting information about the sensory characteristics of food, and (ii) to perform screening test for the selection of products or conditions for the design of a large experiment. Moreover, as it has mentioned in this review chapter, the innovative methods have several advantages. They are faster to perform than conventional, since these methodologies are often performed with assessors semi trained and, even, with consumers, therefore, in these studies could be eliminated both selection and training steps. Other derived advantage of these studies performed with consumers is that the description of the products is done with their own vocabulary; consequently, the companies can understand the preferences of the consumers and align them with the characteristics of the new products. Another advantage of innovative methodologies is that they do not require consensus from the panel, the lack of need for consensus between panellists allows a diversity of points of views, which could provide richer information [21].

Overall above are believed as valuable issues in companies which do not have enough time or economic resources to train a panel. It occurs in both small and large industries, even in developing countries. In many cases, the companies work in a big range of products, and then they cannot train a panel for each product.

However, from a statistical point of view, descriptive analysis is more robust than most novel methodologies, which makes it possible to identify small and subtle differences between products [110]. Another disadvantage of many novel methodologies is related to the interpretation of the sensory terms provided by assessors. For example, in Jimenez et al. [46], the untrained assessors' perception of 'firm/firmness' was not well correlated to the trained panel assessment. This finding suggested that there is a limit in the quality of information that can be obtained via consumers' description. Analysis of assessor descriptions in free profiling, flash profile, open-ended questions, or holistic methodologies is in general a time-consuming, labour-intensive and difficult task [2]. Added to the vocabulary complexity, consumers sometimes use hedonics or benefit-related terms, which in principle can be seen as a limitation because it complicates the analysis [112], however, this information could be interesting to relate product characteristics to marketable features and consumer preference.

The selection of an innovative methodology for a particular application depends on the type of assessors, practical issues and the specific objectives of the studies. When working with consumers it would be generally easier to work with simple methodologies such as CATA questions, open-ended questions or pivot© profile. However, if a study is performed with a trained panel, and quick information about the sensory characteristics of food products is needed, the recommended approach would be to apply flash profile, sorting, projective mapping or polarized sensory positioning due to their higher complexity [2].

Holistic methods based on global similarity, such as sorting and projective mapping seem more appropriate when summarized sensory information is needed [5] and between them, the second might be more discriminant than the first but appears to be more difficult to understand and to perform. Additionally, these methods are based on assessors' global perception of the products, which might enable to identify the main attributes responsible for differences in how they perceive the samples. Thus, they have the extra advantage that can be used for cross-country studies with consumers who speak different languages, without translation problems [2]. More recently, Albert et al. [110] concluded that flash profile gave 'more detailed information about the samples characteristics', whereas projective mapping tended to summarise the information. Veinand et al. [112] reported similar results, this is that, flash profile seemed to be more discriminant than projective mapping, and that, projective mapping assessors reported to be less saturated and tired than flash profile assessors. In contrast, projective mapping was more difficult to perform and to explain. According to the authors [83], in free sorting task assessors are asked to concentrate on the global similarities between the products, whereas in flash profile, assessors dissect their perceptions into attributes.

Polarized sensory positioning or pivot© profile is a good option when the aim of the study is to compare new products with known or reference products or when the sensory characteristics of samples evaluated over time are to be compared [2]. Moreover, polarized sensory positioning might be better suited to compare new products with known ones, and pivot© profile to describe a set of products.

Verbal-based methods (as flash profile and CATA questions) seem to provide more detailed and readily interpretable descriptions and seem to be more discriminant than similarity-based methods (as sorting and projective mapping). Between verbal-based methods, flash profile is more time-consuming but it provides ordinal data rather than mere frequencies. Ares et al. [140] compared projective mapping and CATA questions in the context of ideal products characterisation, and reported that projective mapping might be more adapted to characterize ideal products than CATA questions because the first forces assessors to locate their ideal product within the sample space, whereas with CATA questions the ideal product might be located outside this space. However, from a practical point of view CATA question is perceived as less difficult than projective mapping and is, in general, less time-consuming (about 10 vs. 20 min on the average).

Regarding the time needed by assessors to complete the task, intensity scales, CATA questions, open-ended questions and pivot profile are usually less time-consuming than projective mapping, free choice profiling, flash profile, and polarized sensory positioning. Free choice profiling, flash profile, free sorting and projective mapping require that assessors evaluate all products simultaneously in the same session, since comparisons between them are made. Thus, in order to avoid fatigue and adaptation, the number of samples to be evaluated in a single session is limited when compared to other methodologies such as intensity scales, CATA questions or polarized sensory positioning. For this reason, it could be complicated to use the former methodologies when working with products that require careful temperature control or that have intense and persistent sensory characteristics [46, 110].

Finally, it is important to take into account that most of the innovative methodologies for sensory characterization have been used for a relatively short period of time and have been applied in a limited number of applications, even some methods have been performed in only one product (for example, the Ideal Mapping). For this reason, further research on the applicability, reliability and reproducibility of new methods is still strongly needed.

Acknowledgments

The authors wish to thank the Spanish Ministry of Economy and Competitiveness for financial support (AGL2011-28569).

References

[1] Stone, H., and Sidel, J. L. (2004). *Sensory evaluation practices* (3rd Ed.). London: Elservier Academic Press.

[2] Varela, P., and Ares, G. (2012). Sensory profiling, the blurred line between sensory and consumer science. A review of novel methods for product characterization. *Food Research International, 48*, 893-908.

[3] Meilgaard, M. C., Civille, G. G., and Carr, B. T. (2007). Sensory evaluation techniques (4th Ed.), pp. 173, Boca Raton, FL: CRC Press.

[4] Stone, H., and Sidel, J. L. (1993). Sensory evaluation practices (2nd Ed.), London: Elsevier Academic Press, Inc.

[5] Valentin, D., Chollet, S., Lelièvre, M., and Abdi, H. (2012). Quick and dirty but still pretty good: a review of new descriptive methods in food

science. *International Journal of Food Science and Technology, 47*, 1563-1578.

[6] Moussaoui, K. A. and Varela, P. (2010). Exploring consumer product profiling techniques and their linkage to a quantitative descriptive analysis. *Food Quality and Preference. 21*, 1088-1099.

[7] Lawless, H. T., and Heymann, H. (2010). Sensory evaluation of food. Principles and practices. (2nd edition). New York: Springer.

[8] Greenhoff, K., and MacFie, H. J. H. (1994). Preference mapping in practice. In H. J. H. MacFie and D. M. H. Thomson (Eds.), *Measurement of food preferences* (pp. 137-166).Glasgow: Blackie Academic and Professional.

[9] Lee, S.-Y., Luna-Guzman, I., Chang, S., Barrett, D.M. and Guinard, J.-X. (1999). Relating descriptive analysis and instrumental texture data of processed diced tomatoes. *Food Quality and. Preference, 10*, 447–455.

[10] Cairncross, S. E., and Sjöstrom, L. B. (1950). Flavour profiles: a new approach to flavour problems. *Food Technology, 4*, 308-311.

[11] Dehlholm, C., Brockhoff, P. B., Meinert, L., Aaslyng, M. D., and Bredie, W. L. P. (2012). Rapid descriptive sensory methods – Comparison of Free Multiple Sorting, Partial Napping, Napping, Flash Profiling and conventional profiling. *Food Quality and Preference, 26*, 267-277.

[12] Brandt, M. A., Skinner, E. Z., and Coleman, J. A. (1963). Texture profile method. *Journal of Food Science, 28*, 404-409.

[13] Stone, H., Sidel, J. L., Oliver, S., Woolsey, A., and Singleton, R. C. (1974). Sensory evaluation by quantitative descriptive analysis. *Food Technology, 28*, 24–33.

[14] Williams, A. A., and Arnol, G. M. (1984). A new approach to sensor analysis of foods and beverages. In J. Adda (Ed.), *Progress in Flavour Research, Proceedings of the 4th Weurman Flavour Research Symposium* (pp. 35-50), Amsterdam: Elservier.

[15] Thomson, D. M. H., and McEwan, J. A. (1988). An Application of the repertory grid method to investigate consumer perceptions of foods. *Appetite, 10*, 181-193.

[16] Cliff, M., and Heymann, H. (1993). Development and use of time-intensity methodology for sensory evaluation: A review. *Food Research International, 26*, 375-385.

[17] Perrin, L., Symoneaux, R., Maître, I., Asselin, C., Jourjon, F., and Pagès, J. (2008). Comparison of three sensory methods for use with the

Napping® procedure: Case of ten wines from Loire Valley. *Food Quality and Preference, 19*, 1–11.

[18] Murray, J.M., Delahunty, D. M., and Baxter, I. A. (2001). Descriptive sensory analysis: past, present and future. *Food Research International, 34*, 461-471.

[19] Lawless, H. T., Sheng, N., and Knoops, S. S. C. P. (1995). Multidimensional scaling of sorting data applied to cheese perception. *Food Quality and Preference, 6*, 91-98.

[20] Schiffman, S. S., Reynolds, M. L., and Young, F. W. (1981). Introduction to multidimensional scaling. New York: Academic Press.

[21] Dairou, V., and Sieffermann, J.-M. (2002). A comparison of 14 jams characterized by conventional profile and a quick original method, flash profile. *Journal of Food Science, 67*, 826–834.

[22] Pagès, J. (2005). Collection and analysis of perceived product inter-distances using multiple factor analysis: Application to the study of 10 white wines from the Loire Valley. *Food Quality and Preference, 16*, 642–649.

[23] Risvik, E., McEvan, J. A., Colwill, J. S., Rogers, R., and Lyon, D. H. (1994). Projective mapping: A tool for sensory analysis and consumer research. *Food Quality and Preference, 5*, 263–269.

[24] Adams, J., Williams, A., Lancaster, B., and Foley, M. (2007). Advantages and uses of check-all-that-apply response compared to traditional scaling of attributes for salty snacks. In *7th Pangborn Sensory Science Symposium* (pp. 12–418), August 16th, Minneapolis, MN, USA.

[25] Teillet, E., Schlich, P., Urbano, C., Cordelle, S., and Guichard, E. (2010). Sensory methodologies and the taste of water. *Food Quality and Preference, 21*, 967–976.

[26] Grygorczyk, A., Lesschaeve, I., Corredig, M., and Duizer, L. (2013). Extraction of consumer texture preferences for yogurt: Comparison of the preferred attribute elicitation method to conventional profiling. *Food Quality and Preference*, 27, 215-222.

[27] Giacalone, D., Bredie, W. L. P., and Frøst, M. B. (2013). "All-In-One Test" (AI1): A rapid and easily applicable approach to consumer product testing. *Food Quality and Preference, 27*, 108-119.

[28] Worch, T., Lê, S., Punter, P., and Pagès, J. (2012). Construction of an Ideal Map (IdMap) based on the ideal profiles obtained directly from consumers. *Food Quality and Preference, 26*, 93-104.

[29] Faye, P., Bremaud, D., Teillet, E., Courcoux, P., Giboreau, A., and Nicod, H. (2006). An alternative to external preference mapping based

on consumer perceptive mapping. *Food Quality and Preference, 17*, 604–614.

[30] Parente, m. E., Ares, G., and Manzoni, A. V. (2010). Application of two consumer profiling techiniques to cosmetic emulsions. *Journal of Sensory Studies, 25*, 685-705.

[31] Husson, F., Le Dien, S., and Pagès, J. (2001). Which value can be granted to sensory profiles given by consumers? Methodology and results. *Food Quality and Preference, 12*, 291–296.

[32] Worch, T., Lê, S., and Punter, P. (2010). How reliable are the consumers? Comparison of sensory profiles from consumers and experts. *Food Quality and Preference, 21*, 309–318.

[33] Stampanoni, C. R. (1993). Quantitative flavour profiling: An effective tool in flavour perception. *Food and Marketing Technology.*

[34] Stampanoni, C. R. (1993). The quantitative profiling technique. *Perfumer Flavourist, 18*, 19–24.

[35] Piggott, J. R., Simpson, S. J., and Williams, A. R. (1998). Sensory analysis. *International Journal of Food Science and Technology, 33*, 7-18.

[36] Skinner, E. Z. (1988). The texture profile method. In H. R. Moskowitz (Ed.), *Applied Sensory Analysis of Foods*, (pp. 89-107), Boca Raton, FL: CRC Press.

[37] Szczesniak, A. S. (1963). Classification of textural characteristics. *Journal of Food Science, 28*, 385-389.

[38] Civille, G. V., and Szczesniak, A. S. (1973). Guidelines to training a texture profile panel. *Journal of Texture Studies, 4*, 204-223.

[39] Civille, G. V., and Liska, I. H. (1975). Modifications and applications to foods of the General Foods sensory texture profile technique. *Journal of Texture Studies, 6*, 19-31.

[40] Lawless, H. T., and Heymann, H. (1998). *Sensory evaluation of food: principals and practices*. New York: Chapman and Hall.

[41] Otremba, M. M., Dikeman, M. E., Milliken, G. A., Stroda, S. L., Chambers IV, E., and Chambers, D. (2000). Interrelationships between descriptive texture profile panel sensory panel and descriptive attribute sensory panel evaluations of beef longissimus and semitendinosus muscles. *Meat Science, 54*, 325-332.

[42] Sawyer, F. M., Stone, H., Abplanalp, H., and Stewart, G. F. (1962). Repeatability estimates in sensory panel selection. *Journal of Food Science, 27*, 386–393.

[43] ASTM. (1992). Quantitative descriptive analysis (QDA). *ASTM digital library*. doi: 10.1520/MNL10523M.

[44] Alvarez, M. D., Fernández, C., Jiménez M. J., and Canet, W. (2011). Texture of extra virgin olive oil-enriched mashed potatoes: sensory, instrumental and structural relationships. *Journal of Texture Studies*, *42*, 413-429.

[45] Fernández, C., Alvarez, M. D., and Canet, W. (2008). Steady shear and yield stress of fresh and frozen/thawed mashed potatoes: effect of biopolymers addition. *Food Hydrocolloids*, *22*, 1381-1395.

[46] Jiménez M. J., Canet, W., and Alvarez, M. D. (2013).Sensory description of potato puree enriched with individual functional ingredients and their blends. *Journal of Texture Studies*, in press.

[47] Muñoz, A. M., and Civille, G. V. (1992). The Spectrum descriptive analysis method. In R. C. Hootman (Ed.), *ASTM Manual Series MNL 13, Manual on Descriptive Analysis Testing* (pp. 22-34), West Conshohocken, PA: ASTM International.

[48] Muñoz, A. M., and Civille, G. V. (1998). Universal, product and attribute scaling and the development of common lexicons in descriptive analysis. *Journal of Sensory Studies, 13*, 57-75.

[49] Muñoz, A. M., Civille, G. V., and Carr, B. T. (1992). *Sensory Evaluation in Quality Control*, New York: Chapman and Hall.

[50] Gower, J. C. (1975). Generalised Procrustes analysis. *Psychometrika, 40*, 33-50.

[51] Marshall, R. J., and Kirby, S. P. (1988). Sensory measurement of food texture by free-choice profiling. *Journal of Sensory Studies, 3*, 63-80.

[52] Guy, C., Piggott, J. R., and Marie, S. (1989). Consumer free-choice profiling of whisky. In J. R. Piggott, and A. Paterson (Eds.), *Distilled Beverage Flavour: Recent Developments* (pp. 41-55). Chinchester, UK: Ellis Horwood/VCH.

[53] Oreskovich, D. C., Klein, B. P., and Sutherland, J. W. (1991). Procrustes analysis and its applications to free choice and other sensory profiling. In H. T. Lawless, and B. P. Klein (Eds.), *Sensory science theory and applications in foods* (pp. 353-394). New York: Marcel Dekker.

[54] Piggot, J. R., Sheen, M. R., and Guy, C. (1989). Bridging the gap between laboratory flavour research and the consumer. In G. Charalambous (Ed.), *Flavours and off-flavours, Proceeding of the 6th International Falvour Conference* (pp. 543-552). Amsterdam: Elsevier Science.

[55] Cristovam, E., Paterson, A., and Piggott, J. R. (2000). Differentiation of port wines by appearance using a sensory panel: comparing free-choice and conventional profiling. *European Food Research and Technology, 211*, 65-71.
[56] Murray, J. M. (2001). Sensory approaches to cross-cultural markets. In *Proceedings of the 3rd Annual AIFST Conference*, Brisbane, Australia. *Food Australia, 53*, 503-507.
[57] Kelly, G. A. (1955). The psychology of personal construct: A theory of personality. New York: Norton.
[58] Andani, Z., Jaeger, S. R., Wakeling, I., and MacFie, J. J. H. (2001). Mealiness in apples: Toward a multilingual consumer vocabulary. *Journal Food Science, 66,* 872-879.
[59] Russel, C. G., and Cox, D. N. (2004). Understanding middle-aged consumers' perceptions of meat using repertory grid methodology. *Food Quality and Preference, 15*, 317-329.
[60] Gonzalez-Tomas, L., and Costell, E. (2006). Sensory evaluation of vanilla-dairy desserts by repertory grid method and free choice profile. *Journal of Sensory Studies, 21, 20-33.*
[61] Monteleone, E., Raats, M. M., and Mela, D. J. (1997). Perceptions of starchy dishes: Application of the repertory grid method. *Appetite, 28*, 255-265.
[62] Baxter, I. A., Jack, F. R., and Schroder, M. J. A. (1998). The use of Repertory Grid Method to elicit perceptual data from primary school children. *Food Quality and Preference, 9*, 73-80.
[63] Hersleth, M., Berggren, R., Westad, F., and Martens, M. (2005). Perception of bread: A comparison of consumers and trained assessors. *Journal of Food Science, 70*, S95-S101.
[64] Larson-Powers, N., and Pangborn, R. M. (1978). Paired comparison and time-intensity measurement of the sensory properties of beverages and gelatins containing sucrose or synthetic sweeteners. *Journal of Food Science, 43*, 41-46.
[65] Lee III, W. E., and Pangborn, R. M. (1986). Time-intensity: The temporal aspects of sensory perception. *Food Technology, 40*, 71-82.
[66] Overbosch, P. (1986). A theoretical model for perceived intensity in human taste and smell as a function of time. *Chemical Senses, 11*, 315-329.
[67] Overbosch, P., van den Enden, C. J., and Keur, B. M. (1986). An improved method for measuring perceived intensity/time relationships in human taste and smell. *Chemical Senses, 11*, 331-338.

[68] ASTM. (2011b). Standard guide for time-intensity evaluation of sensory attribute. In *ASTM Standard Guide E1909-11*, West Conshohocken, PA: ASTM International

[69] Guinard, J. X., Pangborn, R. M., and Shoemaker, C. F. (1985). Computerized procedure for time-intensity sensory measurements. *Journal of Food Science, 50*, 543-544.

[70] IFT. (1988). Computers tell 'how sweet it is'. *Food Technology, 42*, 98.

[71] Duizer, L. M., Bloom, K., and Findlay, C. J. (1997). Dual-attribute time–intensity sensory evaluation: A new method for temporal measurement of sensory perceptions. *Food Quality and Preference, 8*, 261–269.

[72] Coxon, A. P. M. (1999). *Sorting data: Collection and analysis.* Thousand Oaks, CA: Sage.

[73] Coxon, A. P. M., Davies, P. M., and Jones, C. L. (1986). *Images of social stratification.* London: Sage.

[74] Miller, D. M., Wiley, D. E., and Wolfe, R. G. (1986). Categorization methodology; An approach to the collection and analyses of certain classes of qualitative information. *Multivariate Behavioral Research, 21,* 135-167.

[75] Black, M. B. (1963). On formal ethnographic procedures. *American Anthropologist, 65*, 1347-1351.

[76] Schiffman, S. S., Reynolds, M. L., and Young, F. W. (1981). *Introduction to multidimensional scaling*. New York: Academic Press.

[77] Popper, P., and Heymann, H. (1996). Analyzing differences among products and panelists by multidimensional scaling. In T. Naes, and E. Risvik (Eds.), *Multivariate analysis of data in sensory science* (pp. 159–184). Amsterdam: Elsevier.

[78] Cartier, R., Rytz, A., Lecomte, A., Poblete, E., Krystlik, J., Belin, E., et al. (2006). Sorting procedure as an alternative to quantitative descriptive analysis to obtain a product sensory map. *Food Quality and Preference, 17*, 562–571.

[79] Tang, C., and Heymann, H. (1999). Multidimensional sorting, similarity scaling and free choice profiling of grape jellies. *Journal of Sensory Studies, 17*, 493–509.

[80] Saint-Eve, A., Pac¸ iKora, E., and Martin, N. (2004). Impact of the olfactory quality and chemical complexity of the flavouring agent on the texture of low fat stirred yogurts assessed by three different sensory methodologies. *Food Quality and Preference, 15*, 655–668.

[81] Faye, P., Bremaud, D., Durand-Daubin, D., Courcoux, P., Giboreau, A., and Nicod, A. (2004). Perceptive free sorting and verbalization tasks

with naive subjects: an alternative to descriptive mappings. *Food Quality and Preference, 15*, 781–791.

[82] Lim, J., and Lawless, H.T. (2005). Qualitative differences of divalent salts: multi-dimensional scaling and cluster analysis. *Chemical Senses, 30*, 719–726.

[83] Blancher, G., Chollet, S., Kesteloot, R., Nguyen, D., Cuvelier, G., and Sieffermann, J.-M. (2007). French and Vietnamese: how do they describe texture characteristics of the same food? A case study with jellies. *Food Quality and Preference, 18*, 560–575.

[84] Lelièvre, M., Chollet, S., Abdi, H., and Valentin, D. (2008). What is the validity of the sorting task for describing beers? A study using trained and untrained assessors. *Food Quality and Preference, 19*, 697–703.

[85] Bécue-Bertaut, M., and Lê, S. (2011). Analysis of multilingual labeled sorting tasks: application to a cross-cultural study in wine industry. *Journal of Sensory Studies*, 26, 299–310.

[86] Chollet, S., Lelièvre, Abdi, H., and Valentin, D. (2011). Sort and beer: Everything you wanted to know about the sorting task but did not dare to ask. *Food Quality and Preference, 22*, 507–520.

[87] Cadoret, M., Lê, S., and Pagès, J. (2009). A factorial approach for sorting task data (FAST). *Food Quality and Preference, 20*, 410–417.

[88] Ares, G., Varela, P., Rado, G., and Gimenez, A. (2011). Are consumer profiling techniques equivalent for some product categories? The case of orange-flavoured powdered drinks. *International Journal of Food Science and Technology, 46*, 1600–1608.

[89] Falahee, M., and MacRae, A. W. (1997). Perceptual variation among drinking waters: The reliability of sorting and ranking data for multidimensional scaling. *Food Quality and Preference, 8*, 389–394.

[90] Abdi, H., Valentin, D., Chrea, C., and Chollet, S. (2007). Analyzing assessors and products in sorting tasks: DISTATIS, theory and application. *Food Quality and Preference, 18*, 627–640.

[91] Abdi, H., Valentin, D., Chollet, S., Chrea, C., Abdi, H., Valentin, D., et al. (2007). Analyzing assessors and products in sorting tasks: DISTATIS, theory and applications. *Food Quality and Preference, 18*, 627–640 (Food Quality and Preference, 20, 410–417).

[92] Ballester, J., Abdi, H., Langlois, J., Peyron, D., and Valentin, D. (2009). The odors of colors: can wine expert or novices distinguish the odors of white, red, and rosé wines? *Chemosensory Perception, 2*, 203–213.

[93] Parr, W.V., Valentin, D., Green, J.A., and Dacremont, C. (2010). Evaluation of French and New Zealand Sauvignon wines by experienced French wine assessors. *Food Quality and Preference, 21*, 56–64.

[94] Rao, V.R., and Katz, R. (1971). Alternative multidimensional scaling methods for large stimulus sets. *Journal of Marketing Research, 8*, 488–494.

[95] Kirkland, J., Bimler, D., Drawneek, A., McKim, M., and Schölmerich, A. (2000). A quantum leap in the analyses and interpretation of attachment sort items. In B. Vaughn, E. Waters, G. Posada, and D. Teti (Eds.), *Patterns of Secure Base Behavior: Q-Sort Perspectives on Attachment and Caregiving in Infancy and Childhood*. Hillsdale, NJ: Lawrence Erlbaum.

[96] Courcoux, P., Qannari, E. M., Taylor, Y., Buck, D., and Greenhoff, K. (2012). Taxonomic free sorting. *Food Quality and Preference, 23*, 30–35.

[97] Santosa, M., Abdi, H., and Guinard, J. X. (2010). A modified sorting task to investigate consumer perceptions of extra virgin olive oils. *Food Quality and Preference, 21*, 881–892.

[98] Cadoret, M., Leˆ, S., and Pagès, J. (2011). Statistical analysis of hierarchical sorting data. *Journal of Sensory Studies, 26*, 96–105.

[99] Bijmolt, T., and Wedel, M. (1995). The effects of alternative methods of collecting similarity data for multidimensional scaling. *International Journal of Research in Marketing, 12*, 363–371.

[100] Heymann, H. (1994). A comparison of free choice profiling and multidimentional scaling of vanilla samples. *Journal of Sensory Studies, 9*, 445–453.

[101] Chollet, S., and Valentin, D. (2001). Impact of training on beer flavor perception and description: are trained and untrained assessors really different? *Journal of Sensory Studies, 16*, 601–618.

[102] Lelièvre, M., Chollet, S., Abdi, H., and Valentin, D. (2009). Beer trained and untrained assessors rely more on vision than on taste when they categorize beers. *Chemosensory Perception, 2*, 143–153.

[103] Piombino, P., Nicklaus, S., LeFur, Y., Moio, L., and Le Quéré, J. (2004). Selection of products presenting given flavor characteristics: an application to wine. *American Journal of Enology and Viticulture, 55*, 27–34.

[104] Derndorfer, E., and Baierl, A. (2006). Development of an aroma map of spices by multidimensional scaling. *Journal of Herbs, Spices and Medicinal Plants, 12*, 39–50.

[105] Deegan, K.C., Koivisto, L., Nakkila, J., Hyvonen, L., and Tuorila, H. (2010). Application of a sorting procedure to greenhouse-grown cucumbers and tomatoes. *LWT-Food Science and Technology, 43*, 393–400.

[106] Delarue, J., and Sieffermann, J.-M. (2004). Sensory mapping using Flash profile. Comparison with a conventional descriptive method for the evaluation of the flavour of fruit dairy products. *Food Quality and Preference, 15*, 383–392.

[107] Lassoued, N., Delarue, J., Launay, B., and Michon, C. (2008). Baked product texture: Correlations between instrumental and sensory characterization using flash profile. *Journal of Cereal Science, 48*, 133–143.

[108] Ferrage, A., Nicod, L., and Varela, P. (2010). How does assessors' performance influence the outcome of alternative sensory methodologies? *Eurosense 2010 - A sense of quality. Vitoria-Gasteiz, Spain, 5–8 September.*

[109] Mazzucheli, R., and Guinard, J. X. (1999). Comparison ofmonadic and simultaneous sample presentation modes in a descriptive analysis of milk chocolate. *Journal of Sensory Studies, 14*, 235–248.

[110] Albert, A., Varela, P., Salvador, A., Hough, G., and Fiszman, S. (2011). Overcoming the issues in the sensory description of hot served food with a complex texture. Application of QDA®, flash profiling and projective mapping using panels with different degrees of training. *Food Quality and Preference, 22*, 463–473.

[111] Tarea, S., Cuvelier, G., and Siefffermann, J.-M. (2007). Sensory evaluation of the texture of 49 commercial apple and pear purees. *Journal of Food Quality, 30*, 1121–1131.

[112] Veinand, B., Godefroy, C., Adam, C., and Delarue, J. (2011). Highlight of important product characteristics for consumers. Comparison of three sensory descriptive methods performed by consumers. *Food Quality and Preference, 22*, 474–485.

[113] Bragato Richter, V., Avancini de Almaido, T. C., Prudencio, S. H., and De Toledo Benassi, M. (2010). Proposing a ranking descriptive sensory method. *Food Quality and Preference, 21*, 611–620.

[114] Rodrigue, N., Guillet, M., Fortin, J., and Martin, J. F. (2000). Comparing information obtained from ranking and descriptive tests of four sweet corn products. *Food Quality and Preference, 11*, 47–54.

[115] Delarue, J., and Loescher, E. (2004). Dynamics of food preferences: a case study with chewing gums. *Food Quality and Preference, 15*, 771–779.

[116] Gómez Alvarado, T., Hernández Cervantes, M., López Velázquez, J., Santiago Cabreral, R., Ramón Canul, L. G., and Juárez Barrientos, J. M. (2010). Caracterización sensorial del queso fresco ''cuajada'' en tres localidades de Oaxaca, México: diferencias en la percepción sensorial. *Revista Venezolana de Ciencia y Tecnología de Alimentos, 1*, 127–140.

[117] Pagès, J. (2003). Recueil direct de distances sensorielles: Application à l'évaluation de dix vins blancs du Val-de-Loire. *Sciences des Aliments, 23*, 679–688.

[118] Ares, G., Deliza, R., Barreiro, C., Giménez, A., and Gámbaro, A. (2010). Comparison of two sensory profiling techniques based on consumer perception. *Food Quality and Preference, 21*, 417–426.

[119] Nestrud, M., and Lawless, H. (2008). Perceptual mapping of citrus juices using projective mapping and profiling data from culinary professionals and consumers. *Food Quality and Preference, 19*, 431–438.

[120] Pagès, J., and Husson, F. (2001). Inter-laboratory comparison of sensory profiles: Methodology and results. *Food Quality and Preference, 12*, 297-309.

[121] Pagès, J., Cadoret, M., and Lê, S. (2010). The sorted napping: A new holistic approach in sensory evaluation. *Journal of Sensory Studies, 25*, 637–658.

[122] Pfeiffer, J., and Gilbert, C. (2008). Napping by modality: A happy medium between analytic and holistic approaches. *Oral communication. Sensometrics.*

[123] Carrillo, E., Varela, P., and Fiszman, S. (2012). Packaging information as a modulator of consumers' perception of enriched and reduced-calorie biscuits in tasting and non-tasting tests. *Food Quality and Preference, 25*, 105–115.

[124] Risvik, E., McEwan, J. A., and Rodbotten, M. (1997). Evaluation of sensory profiling and projective mapping data. *Food Quality and Preference, 8*, 63–71.

[125] Kennedy, J., and Heymann, H. (2009). Projective mapping and descriptive analysis of milk and dark chocolates. *Journal of Sensory Studies, 24*, 220–233.

[126] King, M. C., Cliff, M. A., and Hall, J. W. (1998). Comparison of projective mapping and sorting data collection and multivariate

methodologies for identification of similarity-of-use snack bars. *Journal of Sensory Studies, 13*, 347–358.

[127] Kennedy, J. (2010). Evaluation of replicated projective mapping of granola bars. *Journal of Sensory Studies, 25*, 672–684.

[128] Bárcenas, P., Pérez Elortondo, F. J., and Albisu, M. (2004). Projective mapping in sensory analysis of ewes milk cheeses: a study on consumers and trained panel performance. *Food Research International, 37*, 723–729.

[129] Perrin, L., and Pagès, J. (2009). Construction of a product space from the Ultra-flash profiling method: application to ten red wines from the Loire Valley. *Journal of Sensory Studies, 24*, 372–395.

[130] Nestrud, M. A., and Lawless, H. T. (2010). Perceptual mapping of apples and cheeses using projective mapping and sorting. *Journal of Sensory Studies, 25*, 390–405.

[131] Coomb, C. H. (1964). *The Theory of Data*. New York: Wiley.

[132] Lancaster, B., and Foley, M.(2007). Determining statistical significance for choose-all that-apply question responses. In, *7th Pangborn Sensory Science Symposium*, August 16th, Minneapolis, MN, USA.

[133] Rasinski, K. A., Mingay, D., and Bradburn, N. M. (1994). Do respondents really mark all that apply on self-administered questions? *Public Opinion Quarterly, 58*, 400–408.

[134] Dooley, L., Lee, Y.-S., and Meullenet, J.-F. (2010). The application of check-all-that-apply (CATA) consumer profiling to preference mapping of vanilla ice cream and its comparison to classical external preference mapping. *Food Quality and Preference, 21*, 394–401.

[135] Ares, G., Giménez, A., Barreiro, C., and Gámbaro, A. (2010). Use of an open-ended question to identify drivers of liking of milk desserts. Comparison with preference mapping techniques. *Food Quality and Preference, 21*, 286–294.

[136] Symoneaux, R., Galmarini, M. V., and Mehinagic, E. (2012). Comment analysis of consumer's likes and dislikes as an alternative tool to preference mapping. A case study on apples. *Food Quality and Preference, 24*, 59–66.

[137] ten Kleij, F., and Musters, P. A. D. (2003). Text analysis of open-ended survey responses: A complementary method to preference mapping. *Food Quality and Preference, 14*, 43–52.

[138] Ares, G., Barreiro, C., Deliza, R., Giménez, A., and Gámbaro, A. (2010). Application of a check-all-that-apply question to the development of chocolate milk desserts. *Journal of Sensory Studies, 25*, 67–86.

[139] Parente, M. E., Manzoni, A. V., and Ares, G. (2011). External preference mapping of commercial antiaging creams based on consumers' responses to a check-all-that-apply question. *Journal of Sensory Studies, 26*, 158–166.

[140] Ares, G., Varela, P., Rado, G., and Giménez, A. (2011). Identifying ideal products using three different consumer profiling methodologies. Comparison with external preference mapping. *Food Quality and Preference, 22*, 581–591.

[141] Popper, R., Abdi, H., Williams, A., and Kroll, B. J. (2011). Multi-block *Hellinger* analysis for creating perceptual maps from check-all-that-apply questions. In *9th Pangborn Sensory Science Symposium*, September 7th, Toronto, CA.

[142] Abdi, H., Williams, L. J., Valentin, D., and Bennani-Dosse, M. (2012). STATIS and DISTATIS: optimum multi-table principal component analysis and three way metric multidimensional scaling.Wiley Interdisciplinary Reviews: Computational Statistics, 4, 124–167.

[143] Lado, J., Vicente, E., Manzsioni, A., and Ares, G. (2010). Application of a check-all-that-apply question for the evaluation of strawberry cultivars from a breeding program. *Journal of the Science of Food and Agriculture, 90*, 2268–2275.

[144] Bruzzone, F., Ares, G., and Giménez, A. (2012). Consumers' texture perception of milk desserts II- Comparison with trained assessors' data. *Journal of Texture Studies, 43*, 214–226.

[145] Plaehn, D. (2012). CATA penalty/reward. *Food Quality and Preference, 24*, 141–152.

[146] McCloskey, L. P., Sylvan, M., and Arrhenius, S. P. (1996). Descriptive analysis for wine quality experts determining appellations by Chardonnay wine aroma. *Journal of Sensory Studies, 11*, 49–67.

[147] Chollet, S,. and Valentin, D. (2000). Le degre´ d'expertise a-t-il une influence sur la perception olfactive? Quelques e´ le´ments de re´ponse dans le domaine du vin. *L'Anne´e Psychologique, 100*, 11–36.

[148] Smyth, J. D., Dillman, D. A., Christian, L. M., and Stern, M. J. (2006). Comparing check-all and forced-choice question formats in web surveys. *Public Opinion Quarterly, 70*, 66–77.

[149] Sudman, S., and Bradburn, M. B. (1982). *Asking questions*. San Francisco: Jossey-Bass.

[150] Krosnick, J. A. (1999). Survey research. *Annual Review of Psychology*, 50, 537–567.

[151] Campo, E., Ballester, J., Langlois, J., Dacremont, C., and Valentin, D. (2010). Comparison of conventional descriptive analysis and a citation frequency-based descriptive method for odor profiling: An application to Burgundy Pinot noir wines. *Food Quality and Preference, 21*, 44–55.

[152] Nicod, L., Varela, P., and Ferrage, A. (2010). Could too much training ever be a limitation in products' characterization? *Eurosense 2010 - A sense of quality. Vitoria-Gasteiz, Spain, 5–8 September.*

[153] Ares, G., Bruzzone, F., and Giménez, A. (2011). Is a consumer panel able to reliably evaluate the texture of dairy desserts using unstructured intensity scales? Evaluation of global and individual performance. *Journal of Sensory Studies, 26*, 363–370.

[154] Modell, S. (2005). Triangulation between case study and survey methods in management accounting research: An assessment of validity implications. *Management Accounting Research, 16*, 231–254.

[155] Thuillier, B. (2007). Rôle du CO_2 dans l'Appréciation Organoleptique des Champagnes— Expérimentation et Apports Méthodologiques. Thèse de l'URCA, Reims, France. http://www.abt-sensory-analysis.com/docs/methode_profil_pivot.pdf.

[156] Ares, F., Arrate, E., De Leon, T., Ares, G., and Gambaro, A. (2012). Development of functional milk desserts enriched with resistant starch based on consumers' perception. *Food Science and Technology International, 18*, 465-475.

[157] Busing, F. M. T. A., Groenen, P. J. F., and Heiser, W. J. (2005). Avoiding degeneracy in multidimensional unfolding by penalizing on the coefficient of variation. *Psychometrika, 70*, 49–76.

[158] Poirson, E., Petiot, J.-F., and Richard, F. (2010). A method for perceptual evaluation of products by naive subjects: Application to car engine sounds. *International Journal of Industrial Ergonomics, 40*, 504–516.

[159] Martens, H., Anderssen, E., Flatberg, A., Gidskehaug, L. H., Høy, M., Westad, F., et al. (2005). Regression of a data matrix on descriptors of both its rows and of its columns via latent variables: L-PLSR. *Computational Statistics and Data Analysis, 48*, 103–123.

[160] Popper, R., Rosenstock, W., Schraidt, M., and Kroll, B. J. (2004). The effect of attribute questions on overall liking ratings. *Food Quality and Preference, 15*, 853–858.

[161] Worch, T., Lê, S., Punter, P., and Pagès, J. (2012). Extension of the consistency of the data obtained with the Ideal Profile Method: Would

the ideal products be more liked than the tested products?. *Food Quality and Preference, 26*, 74-80.

[162] Mac Fie, H.J., Bratchell, N., Greenhoff, K., and Vallis, L. V. (1989). Desings to balance the effect of order of presentation and first-order carry-over effects in hall tests. *Journal of Sensory Studies, 4*, 129-148.

[163] Van Trijp, H. C. M., Punter, P. H., Mickartz, F., and Kruithof, L. (2007). The quest for the ideal product: Comparing different methods and approaches. *Food Quality and Preference*, 18, 729-740.

[164] Escofier, B., and Pagès, J. (2008). *Analyses factorielles simples et multiples* (4ème èd.). Dunod, París.

[165] Husson, F., and Pagès, J. (2003). Nuage plan d'individus et variables supplémentaires. *Revue de Statistique Appliquée, 51*, 83-93.

[166] Worch, T., Dooley, L., Meullenet, J. F., and Punter, P. H. (2010). Comparison of PLS dummy variables and fishbone method to determine optimal product characteristics from ideal profiles. *Food Quality and Preference, 21*, 1077-1087.

[167] Worch, T., Lê, S., and Punter, P. (2010). How reliable are the consumers? Comparison of sensory profiles from consumers and experts. *Food Quality and Preference, 21*, 309-318.

[168] Anon. (1999). Taste buds still in control. *Food Engineering, 71*, 14.

[169] Blancher, G., Clavier, B., Egoroff, C., Duineveld, K., and Parcon, J. (2012). A method to investigate the stability of a sorting map. *Food Quality and Preference, 23*, 36-43.

In: New Developments in Sensory Processing … ISBN: 978-1-62808-395-8
Editors: M. Voisin and R. Brunel

Chapter II

Cellular Principles of Serotonergic Modulation: What Do We Learn from Invertebrates?

Citlali Trueta[1], Leonardo Rodríguez-Sosa[2*], Francisco F. De-Miguel[3], Bruno Méndez[3], Carolina Leon-Pinzon[3], Montserrat G. Cercós[1], Víctor Anaya[2] and Gabina Calderón-Rosete[2]

[1]Departamento de Neurofisiología, Subdirección de Investigaciones en Neurociencias, Instituto Nacional de Psiquiatría Ramón de la Fuente Muñiz

[2]Departamento de Fisiología, Facultad de Medicina, UNAM. Ciudad Universitaria, México, D.F.

[3]Instituto de Fisiología Celular-Neurociencias, UNAM. México, D.F.

[*] Correspondence: (L. Rodríguez-Sosa). E-mail: lrsosa@unam.mx.

Abstract

By using small numbers of neurons, serotonergic systems in all animals modulate neural circuits from sensory inputs to motor outputs in an evolutionarily-conserved manner. While in mammals and, in general, in vertebrates the complexity of the nervous system makes it difficult to trace the effects of a neurotransmitter along whole neural circuits, the large and identifiable neurons of invertebrates allow cell by cell studies of the circuits that control stereotyped behaviors and of how serotonin modulates the performance of these circuits. Invertebrates such as decapod crustaceans, leeches and flies have allowed to unveil how serotonin regulates sensory inputs and their associated behaviors, including learning, social, feeding and circadian locomotory behaviors.

While serotonin released as transmitter from presynaptic endings acts in a milisecond time scale in hard-wired circuits, many of the effects of serotonin are slow and long lasting, characteristic of paracrine modulation and are presumably produced through the extrasynaptic release of serotonin from the soma, dendrites and axon varicosities. In particular, somatic release takes place over minutes following excitation and its effects may influence neuronal populations for long periods. The mechanisms of serotonin release from each neuronal compartment have been studied in leech neurons and later confirmed in mammalian neurons. Leech Retzius neurons receive inputs from interneurons and from sensory neurons. While interneurons drive low frequency firing that depends on the degree of electrical coupling between serotonergic neurons and evokes synaptic serotonin release, activation of pressure sensory neurons drive Retzius neurons to fire at higher frequencies that evoke somatic serotonin release. This chapter has as main goals: 1) to provide an overview about serotonergic neuromodulation of behavior from sensory inputs to motor outputs in crustaceans, leeches and flies, and 2) to review the cellular mechanisms that regulate serotonin release from synapses or from extrasynaptic sites, with particular emphasis on the neuronal cell body.

Keywords: Somatic exocytosis, synapse, neuronal functions, modulation, serotonin, 5-HT

Introduction

Increasing advances in Neurosciences in recent years occur at remarkably different levels. On one hand, our understanding of the molecular basis of cell function is rapidly progressing with the increasing number of genes cloned from the nervous system and the study of the role they have in development, neural processing and behavior. On the other hand, the Neurosciences grow at the same rate toward the pole of complex systems, and indeed no-one is as complex as the human brain. It seems feasible that in the next few decades we could be able to have full explanations of several behaviors in terms of neuronal interactions. However, if we consider that the human brain has a hundred billion neurons and it is not accessible to direct experimentation, we can appreciate the magnitude of the task that awaits the neuroscientist.

This makes it attractive to explore the cellular basis of behavior in invertebrate animals, which have small numbers of neurons but yet generate complex behavioral patterns. Such is the case of the crayfish, lobster and leech, in which extensive studies on the neural basis of behavior have generated understanding as to the principles of behavioral integration and on the other hand the accessibility of their large neurons and the possibility of isolating them has permitted to study the fine mechanism of neurotransmitter release. The behavioral repertoire of these animals includes tasks as elementary as nourishment or reproduction and as complex as the establishment of social hierarchies that require coordination between sensory systems, decision-making circuits and motor outputs. In addition to having numbers of neurons that compared to mammalian nervous systems appear finite (e.g. 400 in a leech ganglion), invertebrate nervous systems have the advantage that their neuronal somata have large sizes and stereotyped locations, which makes them identifiable by eye and easily accessible for electrophysiological recording. Moreover, the circuits that produce and regulate specific behaviors, which in humans involve thousands of neurons, in invertebrates are constituted by only a few neurons and in many cases a single invertebrate neuron exerts functions that in vertebrates need thousands of them. Yet, considering their phylogenetic age of several million years, invertebrate behavioral control is highly efficient for survival in harsh environments. Neuronal circuits involved in the regulation of a variety of behaviors can be studied cell by cell in these organisms. Some of the general principles of the neural organization in these animals may still be operating on the complex networks of the human brain (for review see ref. 1). For example, the roles of some neuromodulators like serotonin or dopamine have been conserved throughout the phylogenetic scale [2] and thus, studying the role of these substances and the mechanisms for their release in invertebrate nervous systems has provided information that is also applicable to our own nervous system.

Serotonin Modulation of Behavior

Serotonin (5-hydroxytryptamine, 5-HT) modulates multiple functions and behaviors that are conserved along the zoological scale. For example, aggressive behavior and the establishment of social hierarchies are common to species as distant as primates and crustaceans [3–5]. In vertebrates and invertebrates 5-HT regulates also sexual and feeding behaviors [6,7], and the production of rhythmic motor patterns such as respiration, chewing, locomotion [8–11], digestion [12], and circadian rhythms [13–16]. A distinctive feature of serotonergic systems is that, despite their importance, in all animals they are constituted by very small numbers of neurons. In the leech, for example, there are seven serotonergic [out of 400 total] neurons in each segmental ganglion and in rodents, there are around 9,000 out of 10^{12} total neurons.

These neurons, however, innervate vast areas of practically all the nervous systems. Regulation of serotonin release therefore must be finely regulated and possibly release from different sites is regulated independently.

In spite of the formation of serotonergic synapses by some of the 5-HT terminals, most of the actions of serotonin are slow and long-lasting. Examples for this can be seen in the modulation of mood states, which affect behavior over timescales of hours, suggesting paracrine effects produced by release at extrasynaptic locations. Thus, serotonin seems to be not only a neurotransmitter at fast direct synapses, but also a slow neuromodulator of whole neuronal circuits.

Alterations in brain serotonin levels in humans cause behavioral and mental disorders such as depression, schizophrenia, anorexia or obsessive-compulsive disorder [17–19].

In spite of the importance of the serotonergic system for the regulation of all these functions and of the availability of serotonergic pharmacological treatments for these diseases, the information available on the mechanisms regulating endogenous release and the modes of operation of these drugs, are scarce. Answers to questions concerning the mechanisms of serotonin release and their regulation have begun to be found in leech serotonergic neurons and confirmed in rodents.

Serotonergic Modulation of Photosensitivity and Circadian Rhythms

A good example as to how serotonin modulates behavior from the sensory inputs is provided by photosensitivity in crustaceans. In crayfish the light sensivity is mediated by retinal photoreceptors but also by photoreceptors in the supraesophageal ganglion (brain photoreceptors, BPR) and in the sixth abdominal ganglion [6th AG; caudal photoreceptors; 20]. Serotonin modulates the activity of all these photoreceptors. Some of the pathways between the retina and the protocerebrum in the supraesophageal ganglion seem to be serotonergic [21] and the eyestalk has 5-HT_{1A} receptors, the amount of which varies during light-dark cycles as shown by autoradiography of bound $^{(3H)}$8-OH-DPAT [22]. In retinal photoreceptors serotonin enhances sensitivity to light [23, 24]. The brain photoreceptors are two clusters of red-brown pigmented cell somata in the anterior margin of the cerebral ganglion that have axonal terminals in the neuropil of the protocerebral bridge together with serotonin-containing structures [25], which participate in circadian rhythms [20, 26]. The caudal photoreceptors, which are one neuron in each side of the 6th AG, respond directly to light with phasic-tonic activity and also respond trans-synaptically to mechanosensory inputs [27]. They participate in locomotor modulation [for review see ref. 28, 29] and their electrical stimulation initiates backward walking [30]. Both mechanical and light sensitivity interact with each other, suggesting that these neurons activate command neurons that trigger walking behavior in crayfish [30–33]. Furthermore, the caudal photoreceptor is part of a circuit partially characterized mediating light-avoidance responses associated with a preference for dark habitats [33–35]. These receptors behave as a circadian oscillator [28] that participates in the synchronization of the locomotor circadian rhythm in crayfish [for review see ref. 26]. In support to this, injury of the connectives between the 6th and 5th abdominal ganglia in *P. clarkii* abolishes backward burrowing behavior in response to bright lights [33] and causes a phase-shift in the locomotion circadian rhythm [29]. Exogenous 5-HT and/or an agonist of the 5-HT_{1A} receptor also produce a phase-shift in the circadian rhythms for spontaneous and light-induced activities of the caudal photoreceptors [36; for review see refs 28, 29]. The 6th AG contains serotonergic neurons that have been identitified by immunostaining. Topical application of serotonin to isolated ganglia kept in the dark increases the spontaneous firing rate of the caudal photoreceptors and reduces its response to light [37].

In animals throughout the phylogenetic scale light-dark cycles regulate the secretion of hormones, which act as an intermediate step between the environmental synchronizing stimuli (also called zeitgeber) and behavior. Crustaceans are good examples where the participation of hormones in circadian rhythms has been studied in detail [28, 35]. Several hormones contribute to the circadian behavior and their release is modulated by serotonin. Serotonin itself circulates in the haemolymph, thus suggesting its role as a neurohormone [38, 39]. Consistent with this, injection of 5-HT into the circulatory system causes the release of hyperglycemic hormone in the crayfish *Orconectes limosus* [40; for review see ref. 41], the red and black pigment dispersing hormones in the fiddler crab *Uca pugilator* [42] and the gonad-stimulating hormone in crayfish *Procambarus clarkii* [43, 44].

The presence, biosynthesis, degradation, and release of 5-HT after stimulation have been reported in the isolated eyestalk of the crayfish [24] and the chromatographic quantification of its precursors, L-tryptophan and 5-hydroxytryptophan, suggest that the biosynthetic pathway of 5-HT in the nervous system of crustaceans conserves the conventional enzymes tryptophan hydroxylase and aromatic L-amino acid decarboxylase [24, 37, 45]. A basal release of 5-HT was detected in isolated eyestalks. The amount recovered was increased 2-3 fold after blocking 5-HT uptake with fluoxetine (1 μM). Incubation of eyestalks in high potassium [KCl 80 mM] solutions released 5-HT. Electrical stimulation of the optic nerve released 5-HT as a function of the intensity of stimulation. Both basal and evoked release were suppressed by lowering the calcium concentration in the medium. Topical application of 5-HT on the isolated eyestalk has different effects on the electrical activity of different secretory cells in the X-Organ [46], the primary neuroendocrine system in the crayfish. Serotonin has an excitatory effect on the largest neurons, which probably contain the hyperglycemic hormone. This effect persists for about 30 min after washout from the bathing solution, but desensitizes upon successive applications of serotonin. A plausible mechanism for this excitatory effect is the reduction of a K^+ current [47]. On the other hand, neurons that produce the red pigment concentrating hormone are inhibited by serotonin [48], through an increase in the K^+ conductance through Ca^{2+}- and voltage-dependent Big Potassium [BK] channels. Both effects of serotonin on the previously mentioned peptidergic neurons were blocked by methysergide, a non-specific antagonist of serotonergic receptors.

These results support the notion of 5-HT as a neurotransmitter or modulator in the crayfish eyestalk.

Serotonin is also involved in the modulation of circadian rhythms in *Drosophila*. For example, agonists of the 5-HT$_2$Dro, a *Drosophila* ortholog of the mammalian 5-HT$_2$ receptor, increase early day activity [49] and 5-HT levels are decreased in flies maintained in constant darkness for long periods [50]. A short pulse of light during the night produces a phase shift in the circadian clock in *Drosophila*. This shift involves the phosphorylation and subsequent degradation of the clock protein timeless TIM [51,52] by glycogen synthase kinase 3 beta [GSK3β] in response to photic signals transmitted by the circadian cryptochrome photoreceptor [53, 54]. Serotonin inhibits these phase shifts in the circadian clock through activation of 5-HT$_{1B}$Dro receptors in clock neurons [50], which decreases the activity of GSK3β, thus increasing the stability of TIM.

During *Drosophila* larval development serotonin signaling is required for proper regulation of larval photobehavior. Dopa decarboxylase (DDC)- or Tryptophan hydroxylase (dTRH)-null mutations cause an increase in the length of the pauses and changes in direction produced by light, and thus reduce the linear movement characteristic of the larval photophobic behavior [55].

Serotonin as a Modulator of Social Behavior and Agression

The role of serotonin modulating aggressive and dominant behavior has been studied in lobster, crayfish and fruit fly. Two lobsters sharing the same space develop an aggressive behavior extending their claws forward and walking erected on the tips of their flexed legs. Both animals undergo agonistic encounters that determine which will dominant and which will be subordinate, depending on their size, strength and fighting capabilities. The dominant individual keeps walking in the erected posture on flexed legs while the subordinate stays in the corners of the tank extending its appendages and flattening its body near the bottom, while avoiding the dominant mate [for review, see ref. 4]. The spontaneous tonic activity of motoneurons and the resistance of the muscles controlling downward and upward movements of the legs in crayfish are larger in dominants than in subordinates [56]. Serotonin and octopamine seem to be responsible for the induction of these opposite behaviors, respectively, since injection of serotonin in lobsters produces the dominant posture with flexed appendages and abdomen, while injection of octopamine induces the subordinate flat posture with extended appendages [38]. A number of 5-HT-immunopositive neurons have been identified in central ganglia of the lobster [57–60], but the endogenous serotonin that modulates these behaviors is released specifically by two pairs of neurons: one located in the fifth thoracic ganglion (T5) and the other in the first abdominal ganglion (A1). The axons of these neurons ramify within the CNS and also send branches to peripheral neurosecretory organs, from which they release serotonin into the haemolymph. Although stimulation of these serotonergic neurons does not activate motoneurons directly, they are activated by command neurons that initiate the flexor pattern [61], and serotonin released from their central terminals in turn enhances the effect of the flexor command neurons by increasing the responsiveness of the motoneurons that excite flexor muscles and the ones that inhibit extensor muscles [for review see ref. 4]. All this occurs though modifying the frequency of activation of the synaptic inputs to the motoneurons and not through direct effects on the motoneuron properties [62]. In addition, serotonin released into the haemolymph increases the response of the neuromuscular junctions by acting pre- and post-synaptically: presynaptically, serotonin increases excitatory transmitter release and postsynaptically it prolongs muscle contractions [63–65]. Thus, serotonin acts as a "gain-setter" for the flexor motor command, amplifying the responses it triggers [for review see ref. 4]. On the other hand, when the command neurons that trigger the motor pattern for appendage extension in the subordinate behavior are activated, the T5 and A1 serotonergic neurons are inhibited. Octopamine contributes to this inhibition and, contrarily to the effects of serotonin, it activates the motoneurons that excite the extensor muscles and the ones that inhibit flexor muscles, and complementarily it inhibits the motoneurons that excite the flexor muscles and the ones that inhibit extensor muscles.

Serotonin also seems to act as a motivational cue for fighting in crayfish and lobsters. Serotonin injected to animals that have already asumed a subordinate role increases the time they remain engaged in fighting with the dominants, presumably increasing their willingness to fight [66].

The serotonergic modulation of behavioral-related circuits in decapods depends on the social status of the animal. For example, in the motoneurons that control the elevation or depression of the legs in crayfish, serotonin increases the amplitude and duration of excitatory postsynaptic potentials (EPSPs) triggered by sensory spikes, the tonic firing rates and the resistance reflex amplitudes in dominant animals but not in subordinates [56]. The modulation of synaptic transmission at connections between mechanosensory afferents from the tail fan and the lateral giant neuron that produces the escape reaction in the crayfish [67, 68] is also social-status-dependent: while serotonin facilitates synaptic transmission in dominant or isolated animals, it reduces transmission in subordinates [69]. This seems to be associated to changes in the type and possibly the density of serotonin receptors expressed by the animals [70, 71]. The modulatory effect of serotonin also depends on the pattern of cell exposure. High or low concentrations and rapid or slow exposures produce opposite modulatory effects on sensory-evoked EPSPs in the lateral giant neurons [71]. In addition, brief exposures produce transient modulatory effects, whereas longer exposures produce long-term facilitation [72, 73]. Interestingly, lateral giant neurons also excite A1 and T5 serotonergic neurons, suggesting a feedback loop in this circuit [74].

The role of serotonin in aggresssive behavior has also been studied in the fruit fly *Drosophila melanogaster*. This organism has been classically used to study mechanisms of gene expression and their role in a variety of functions and in recent years Edward Kravitz's and other groups have developed powerful tools to manipulate the biosynthesis and release of serotonin as well as the expression of 5-HT receptors, to study the role of this amine in a variety of behaviors.

In *Drosophila* as in lobsters, pairs of males establish hierarchical relationships that start by frequent approaching and tapping on the opponent´s leg. As fights escalate, both combatants display intermediate- and high-intensity behavioral patterns such as lunging, where one fly rears up on hind legs and snaps down on the other; holding, where one fly grasps the opponent with forelegs and tries to immobilize it; boxing, where both flies rear up on hind legs and strike the opponent with forelegs; or tussling, where both flies tumble over each other, sometimes leaving the food surface. Once dominance is established, only winners lunge while defeated flies no longer lunge and retreat from the food surface when confronted by winners [75, 76].

Normal adaptive forms of aggression aimed at territorial control and social dominance are related to the function of serotonergic neurons [77, 78]. Increasing 5-HT levels in flies by increasing 5-hydroxytryptophan in the diet or by constitutive over-expression of dTRH in serotonergic neurons increases the aggression [79]. On the other hand, reductions in 5-HT neuron function by feeding flies with inhibitors of serotonin synthesis or by silencing the expression of dTRH in serotonergic neurons, does not affect the fighting frequency.

Transgenic lines have allowed to transiently manipulate serotonin release at different times during development or in adult animals by use of a temperature-sensitive form of dynamin, a protein involved in endocytosis, selectively expressed in serotonergic neurons [80]. Flies with reduced serotonergic neurotransmission had normal levels of overall activity and encounters, but displayed substantial reductions in male mid-intensity aggression, resulting in fewer dominance relationships. By contrast, when serotonergic cells are selectively activated by expressing temperature-dependent dTrpA1 cationic channels, aggression is expressed faster; fights become intensified and continue even after social dominance has been established [80]. These results agree with findings in rats, where the regulation of 5-HT release determines the transition between normal and escalated types of aggression [78].

Serotonin modulates also the fly courtship behavior. Two clusters of 5-HT neurons located in the sexual abdominal ganglion innervate parts of the male reproductive apparatus concerned with seminal fluid and sperm transfer. Fruitless mutants that do not express serotonin showed abnormal behavior including courtship among males [81]. This also is consistent with findings in mammals in which drug-induced 5-HT reductions can induce homosexual behavior of males [82, 83].

Serotonergic Modulation of Feeding Behavior

Serotonin regulates every stage of feeding behavior in the leech, from hunting through satiety. Hungry leeches locate themselves in shallow waters and upon detection of water movements (which can be produced by a potential prey) start swimming toward the source of the waves. Serotonergic neurons play a fundamental role in swim initiation. Swimming in these organisms is coordinated by a series of interneurons with oscillatory activities that rhythmically stimulate and inhibit the motoneurons that excite and inhibit, respectively, the longitudinal ventral and dorsal muscles in an alternated manner. Activation of this central pattern occurs progressively along the CNS ganglion chain, producing ondulations of the body from the cephalic to the caudal ends [84]. The rhythmical activity of these neurons can be recorded from the segmental ganglia or in the connective nerves and is maintained in isolated ganglion chains, which has facilitated the study of these circuits.

Mechanosensory neurons polysynaptically excite several serotonergic neurons, including cells 21, 61 [85] and Retzius neurons [86, 87]. Stimulation of cells 21 and 61 immediately triggers swimming episodes in isolated ganglion chains through the activation of interneurons of the swimming central pattern generator, which in turn gives feedback to the serotonergic neurons [88]. Stimulation of Retzius neurons or application of serotonin in the bath produces swimming episodes with a delay [9], through the modulation of the activity of oscillatory interneurons that trigger swimming. For example, serotonin changes the excitability of one cell, called cell 204, by lowering its firing threshold through the modulation of several sodium channels [89, 90]. Serotonin also modulates the swimming motoneurons by promoting their alternated rhythmical activity [91]. Thus, serotonin seems to be the determining transmitter and modulator for the activation of leech swimming. In fact, the probability of swimming is well correlated with the concentration of serotonin in the blood [9] and in the CNS [92] of the animal. Serotonin released from Retzius neurons has also a neuromodulatory effect on conduction block in the branching points of the dendritic tree of the pressure-touch- and nociceptive-sensory neurons [93]. In this way, serotonin modulates the conduction of sensory inputs in the leech CNS.

Once a leech finds a potential prey, it starts an exploratory phase. During swimming and this exploratory phase, the activity of serotonergic neurons is increased [94]. When the lips feel a warm surface, some serotonergic neurons are activated, firing high-frequency impulse trains, and serotonin activates ingestion, directly stimulating the salivary glands to secrete saliva, stimulating the jaws to bite and the pharynx to produce peristaltic movements, which continue through the ingestion phase.

During ingestion, the body wall muscles are relaxed and the skin secretes an increased amount of mucus. These functions are also regulated by serotonin, which inhibits the production of excitatory synaptic potentials in the muscles through actions on central neurons [95] and directly produces a chloride-dependent hyperpolarization in the muscles [96], resulting in the relaxation of the body wall. Retzius neurons also stimulate mucus production through direct actions on the mucosal glands of the skin [97].

Distention of the body wall after feeding hyperpolarizes the serotonergic neurons and terminates ingestion, producing a satiety state [98]. Administration or elimination of serotonin causes behavioral alterations such as hyperphagia and anorexia respectively [10]. Satiated leeches have a smaller concentration of serotonin in the CNS than hungry leeches and tend to go to deeper water in the ponds and to reduce their swimming.

Serotonin controls feeding behavior also in *Drosophila*. Reduction in dTRH levels by constitutive knockdown in adults results in increased feeding (measured as the number of mouth hook contractions), without any change in circuit architecture. However, when the dTRH synthesis is disrupted during CNS development, feeding is depressed and the serotonergic projections to the gut display increased branching, number and size of varicosities along the neurite length, demonstrating a role for serotonin as a modulator of neuron growth with direct effects on neuronal architecture, and on the function of this neural circuit [99].

Serotonergic Modulation of Sensory-Related Learning and Memory

In addition to the modulation of feeding behavior, serotonin in the leech modulates facilitation, sensitization and dishabituation of the shortening reflex in response to touch in the head [100, 101]. This modulation is exerted on the excitability of a set of neurons connected in series, called S cells, which are essential for learning in the leech [102]. The activity of the S cell network is modulated by serotonin throughout the ganglion chain [103].

Serotonin also regulates olfactory memory in *Drosophila*. This organism has, again, been useful to study this behavior at the molecular level. To elucidate the role of serotonin in aspects of both short- and long-term learning and memory, populations of *Drosophila* were trained by alternately exposing them to pairs of odor cues, with shock or mechanical shaking as negative reinforcement and sucrose as positive reinforcement [104, 105]. Agonists and antagonists of 5-$HT_{1A/1B}$Dro, 5-HT_2Dro and 5-HT_7Dro decreased short- and long-term memory, measured as the difference between the percentage of flies making the correct choice and the percentage of flies making the incorrect choice [106]. These findings were validated using genetic methods in which knockdown of 5-$HT_{1A/1B}$Dro or 5-HT_2Dro receptors also produced significant deficits in the performance index.

Mechanisms of Serotonin Release

The mechanisms by which serotonin is released as transmitter, as a paracrine modulator or as a hormone have been studied in serotonergic Retzius neurons of the leech. The leech central nervous system with its chain of stereotypedly arranged intermediate ganglia, has been one of the most useful preparations to study integrative functions of the central nervous system, from basic cellular principles to behavior and development. Most of the leech central ganglia contain seven serotonergic neurons: two Retzius neurons, which are the largest in the ganglia, a pair of dorso-lateral neurons, also called cells 21 or DL, a pair of ventro-lateral neurons called 61 or VL and a single neuron in the medial posterior packet. The three first ganglia of the CNS have an additional pair of serotonergic neurons in the medial anterior packet.

The large soma of Retzius neurons sends a primary axon that branches and produces mutiple neurites within the neuropile and divides into secondary axons that are sent to the periphery and to anterior and posterior ganglia through the connective nerves. The extensive neuropilar arborizations form about 32 non-rectifying electrical connections with the other Retzius neuron in the ganglion [107–109]. These electrical synapses allow synaptic and action potentials initiated in one cell to propagate across the coupled dendrites to the other Retzius neuron, thus contributing to synchronous firing [109–111]. The rest of the 5-HT neurons in the ganglion branch within the central nervous system, where all are electrically coupled and receive common cholinergic synaptic inputs that induce a synchronic electrical activity [112].

Synaptic Release of Serotonin

Most of our current knowledge about the fine mechanisms of synaptic release of serotonin come from experiments made by John G. Nicholls and co-workers in cultured leech Retzius neurons [For review see ref. 113, 114], which in spite of being adult survive in culture, where they maintain their electrical properties and continue to synthesize and release serotonin upon electrical stimulation [115–117]. Retzius neurons form chemical synapses onto pressure sensory (P) neurons and upon electrical stimulation with intracellular current pulses to produce action potentials or with voltage depolarizations, they release quanta of serotonin in a calcium-dependent manner [116, 118]. Synaptic terminals of cultured Retzius neurons contain clear vesicles apposed to the plasma membrane and surrounded by dense core vesicles [119]. Amperometric records of serotonin release have shown that a quantum is composed by about 4,000 molecules of serotonin and several quanta are released from a single synapse in response to an action potential. Moreover, stimulation with pairs or trains of impulses induces release of serotonin from large (100 nm) perisynaptic dense core vesicles, each of which contain about 67,000 serotonin molecules [120]. Serotonin released from synapses activates autoreceptors, which activate a chloride conductance, thus reducing the amplitude of the action potential arriving at the terminal and therefore decreasing presynaptic firing and further serotonin release [121].

Extrasynaptic Release of Serotonin

Many of the effects of serotonin described above are slow, diffuse and long-lasting. Therefore they are unlikely to be produced by serotonergic synapses, in which transmission occurs in milliseconds. Instead, serotonin in many cases seems to act by bathing whole neuronal circuits through the so-called "volume transmission" [122, 123]. The source of serotonin acting in this manner seems to be the extrasynaptic release from the neuronal somata, dendrites and axonal varicosities. Somatic and dendritic release has been shown for multiple transmitters and peptides [for review see ref 124] in vertebrates and invertebrates, and again, the mechanisms for somatic serotonin release have been studied step by step in leech Retzius neurons. In comparison to synaptic release, which responds to single action potentials or to low frequency firing and occurs in a few dozens of milliseconds [113], somatic release requires trains of action potentials at high frequency and lasts for minutes after stimulation [125]. Here we will describe briefly the general mechanisms for somatic exocytosis. Serotonin in the cell body of Retzius neurons is packed in dense core vesicles [126], which are arranged in clusters and in association with mithochondria and endoplasmic reticulum [127, 128]. These multivesicular clusters are connected to the plasma membrane through microtubule bundles that act as rails for the active transport of the vesicles in response to electrical activity. Brief stimulation with trains of 10 action potentials produced in Retzius neurons by current injection through a microelectrode at 10 or 20 Hz trigger calcium entry through L-type channels [125] and this entry in turn activates calcium-induced calcium release from intracellular stores, most likely endoplasmic reticulum that mediates the space between the plasma membrane and the vesicle clusters and mitochondria [127]. This mechanism has also been found for somatic release in substantia nigra [129] and thalamic neurons [130]. Thus, somatic secretion uses mechanisms that differ from those in synaptic terminals and are similar to those for secretion from excitable endocrine cells. Evidence suggests that arrival of calcium waves at the mitochondria induces the synthesis of ATP and the increase of cytoplasmic ATP then activates the kinesin motors that transport the vesicle clusters towards the plasma membrane at a velocity of 15-95 nm/sec. Incubation of neurons with colchicine, a blocker of microtubule polymerization, abolishes exocytosis, thus showing that active vesicle transport is required as an intermediate step between excitation and exocytosis [131]. The amount of ATP molecules used per vesicle fused, estimated by combining fluorescence imaging of exocytosis, electron microscopy and mathematical modeling, is 0.4-72 molecules and is inversely proportional to the resting distance between the vesicle cluster and the plasma membrane [131], thus suggesting a safety factor that prevents spontaneous mobilization of the vesicles at rest. Moreover, this transport mechanism confers a time-dependence to somatic exocytosis, since vesicle transport occurs at an average of 50 nm/sec and explains the long delay and duration of exocytosis after the onset of electrical activity. Exocytosis ends once all the vesicles in the clusters fused with the membrane and upon endocytosis vesicles are collected into multivesicular bodies, where they are degraded and transported back to the perinuclear regions of the cell. There, the contents of the multivesicular bodies are re-used to form new vesicles and clusters that upon electrical stimulation start their migration towards more external regions of the cell [128]. Our estimates are that up to 100,000 vesicles may release their contents in response to a single 20 Hz train of stimulation. Therefore, this slow exocytosis may allow a gradual increase in the concentration of serotonin circulating in the extracellular space and the amount of serotonin released in this way may suffice to produce the modulatory effects observed in the leech nervous system when extracellular serotonin is applied [see ref. 9].

The use of multiphoton illumination to visualize serotonin in raphe neurons of rodents has shown that serotonin spots respond to high potassium depolarization by mobilizing towards the plasma membrane and releasing their contents to the extracellular space [132, 133]. Although the mechanisms of exocytosis in these neurons seem similar to those in leech Retzius neurons, two aspects remain to be clarified in raphe neurons. The first one is the nature of the organelles that contain serotonin, since the diameter of the fluorescent spots is more consistent with clusters of vesicles or multivesicular bodies than with single vesicles. The second is the nature of the vesicle transport, since the analysis suggests that it could be random, although no direct experimentation has been carried out yet to test for this.

In addition to somatic exocytosis, other areas of the neurons also release serotonin by exocytosis in vertebrates and invertebrates. For example, perisynaptic regions in leech synapses in culture release serotonin from dense core vesicles upon repetitive stimulation [Trueta and De-Miguel, in preparation]. Extrasynaptic release has also been seen from dendrites of vertebrate raphe neurons, and the axons of serotonergic neurons in rodents and leech contain secretory vesicles that seem to fuse in non-synaptic sites [for review, see ref. 124].

Sensory and Non-Sensory Inputs Governing Synaptic and Extrasynaptic Release

A question of relevance for this review concerns the physiological stimuli that govern serotonin release. To study this, the leech central nervous system again provides multiple advantages, since the size of the neurons and their relatively stereotyped sizes and positions in the ganglion allow simultaneous intracellular recordings from pairs of neurons. This approach has shown that Retzius neurons receive inputs from unidentified neurons that connect onto the coupled dendrites of Retzius neurons and also from every single sensory neuron in the ganglion through polysynaptic connections onto non electrically-coupled neurites. The input to the coupled neurites produces synchronic synaptic potentials, which activates tonic firing at low frequencies between 0.1 and 1 Hz. On the other hand the activation of pressure mechanosensory neurons drives the firing frequency of Retzius neurons up to 50 Hz [87]. Since synaptic serotonin release is activated at low frequencies and extrasynaptic release requires high firing frequencies [125], this sensory modulation of the firing frequency of Retzius neurons leads to different modes of serotonin release. In addition, serotonin or activation of Retzius neurons reduces the duration of conduction block in mechanosensory cells [93].

The common chemical (most likely cholinergic) input onto the electrically coupled dendrites of Retzius neurons [110] produces sinchronous synaptic potentials in both coupled Retzius neurons. The amplitude of these chemical synaptic potentials is regulated by an interesting phenomenon: the coupling coefficient of the cells determines the amplitude of the synaptic responses by controlling the impedance of the dendrites [109, 111]. Thus, at low dendritic coupling coefficients the amount of synaptic currents remaining in the dendrites increases, and the synaptic potentials become larger. As the coupling increases, more synaptic current flows to the coupled dendrite and the local synaptic potential becomes smaller. This reduction in the amplitude of synaptic potentials correlates with a reduction in the basal firing frequency of Retzius neurons. The synaptic current that leaks to the coupled neuron contributes to its firing frequency by summing to the local synaptic currents [111]. Therefore, when electrical coupling is low, neurons determine their firing more independently from each other, whereas at high coupling, neurons share their synaptic inputs and their firing becomes more dependent on that of the coupled neuron.

The firing frequency of Retzius neurons is modulated by synaptic inputs from mechanosensory neurons. Each central ganglion in the leech CNS contains six neurons that respond to touch in the skin, four neurons that respond to pressure and four neurons that respond to noxious stimuli. Because of these responses, sensory neurons are known as T (touch), P (pressure) and N (nociceptive) cells respectively [134]. Each neuron of each modality has a characteristic receptive field in the skin [135–137] and stimulation of this receptive field evokes responses that depend on the strength of stimulation, ranging from single impulses to bursts of action potentials. For example, gentle touch to the skin evokes trains of action potentials in the T cell innervating that particular field. These trains adapt rapidly. Increasing the stimulation strength activates P cells, which increase their firing rate as the stimulus strength increases, and at a certain stimulus force, N cells become active [134]. Responses of any of these neurons are followed by responses in Retzius neurons, through the activation of polysynaptic pathways that arrive at the non-coupled dendrites of both Retzius neurons in each ganglion [87]. However, the synaptic load onto Retzius neurons depends on the sensory modality of the different neurons. Intracellular stimulation of sensory neurons with trains of impulses has provided information as to how each sensory modality regulates the firing rate of Retzius cells [87]. Stimulation of T cells produces large synaptic potentials that depress as the train proceeds. Because of this, T cells fail to produce sustained active responses in Retzius neurons and therefore may activate synaptic but no extrasynaptic exocytosis. By contrast, P cells produce polysynaptic responses in Retzius neurons that facilitate upon trains of impulses and skin stimulation with forces that produce firing of P cells also produces the firing or Retzius neurons, which follow the bursting responses of P cells and fire at frequencies up to 50 Hz, enough to evoke somatic serotonin exocytosis. Although the polysynaptic inputs of N cells onto Retzius neurons are similar to those of P cells, the low physiological frequencies at which N cells respond act mostly by reinforcing the activity of P cells, the later being the main drivers of Retzius cell activity [87]. Thus, depending on the sensory input that Retzius neurons receive, they may release serotonin exclusively from synapses or may also recruit exocytosis from other neuronal compartments.

Conclusion

By acting as a neurotransmitter, neuromodulator and neurohormone serotonin regulates a multiplicity of sensory systems and related behaviors, including social and feeding behaviors and circadian rhythms. In invertebrates, which allow studying neuronal circuits at the cellular level, serotonin modulates behavioral circuits at each stage from sensory inputs to motor outputs. Serotonin acts as a transmitter when released from synapses in hard-wired circuits, and also acts as a paracrine or hormonal agent at a distance from the serotonergic neurons when released from extrasynaptic sites in the soma, dendrites and axonal varicosities. Extrasynaptic exocytosis of serotonin uses intracellular mechanisms different from those at synapses, although both mechanisms may coexist in the same neuron. Somatic serotonin release in leech Retzius neurons requires electrical activity at high frequencies, which are evoked by activation of pressure mechanosensory neurons. In turn, serotonin release modulates the electrical activity of pressure mechanosensory neurons. Many of the mechanisms described here may represent general principles at the cellular level that are preserved in higher order animals including humans.

Acknowledgments

This work was funded by PAPIIT-UNAM IN211511 grant to FFM, by CONACYT 43703 and TWAS 03-263 RG/BIO/LA grants to CT and by CONACYT 130031 to FFM, CT and LRS.

References

[1] Clarac F, Pearlstein E. Invertebrate preparations and their contribution to neurobiology in the second half of the 20th century. 2007 Apr;113–61.

[2] Weiger WA. Serotonergic modulation of behaviour: a phylogenetic overview. *Biol Rev Camb Philos Soc.* 1997 Feb;72(1):61–95.

[3] Raleigh MJ, McGuire MT, Brammer GL, Pollack DB, Yuwiler A. Serotonergic mechanisms promote dominance acquisition in adult male vervet monkeys. *Brain Res.* 1991 Sep 20;559(2):181–90.

[4] Kravitz EA. Serotonin and aggression: insights gained from a lobster model system and speculations on the role of amine neurons in a complex behavior. *J. Comp. Physiol.* A. 2000 Mar;186(3):221–38.

[5] Alekseyenko OV, Lee C, Kravitz EA. Targeted manipulation of serotonergic neurotransmission affects the escalation of aggression in adult male Drosophila melanogaster. *PLoS ONE.* 2010;5(5):e10806.

[6] Hull EM, Lorrain DS, Du J, Matuszewich L, Lumley LA, Putnam SK, et al. Hormone-neurotransmitter interactions in the control of sexual behavior. *Behav. Brain Res.* 1999 Nov 1;105(1):105–16.

[7] Goldburt V, Sabban BA, Kleinhaus AL. Serotonin depletion inhibits feeding in carnivorous leeches (*Haemopis*). 1994 Jan;47–53.

[8] Saller CF, Stricker EM. Hyperphagia and increased growth in rats after intraventricular injection of 5,7-dihydroxytryptamine. *Science*. 1976 Apr 23;192(4237):385–7.

[9] Willard AL. Effects of serotonin on the generation of the motor program for swimming by the medicinal leech. *J. Neurosci.* 1981 Sep;1(9):936–44.

[10] Lent CM. Serotonergic modulation of the feeding behavior of the medicinal leech. *Brain Res. Bull.* 1985 Jun;14(6):643–55.

[11] White SR, Fung SJ, Jackson DA, Imel KM. Serotonin, norepinephrine and associated neuropeptides: effects on somatic motoneuron excitability. *Prog. Brain Res.* 1996;107:183–99.

[12] Marder E, Bucher D. Understanding circuit dynamics using the stomatogastric nervous system of lobsters and crabs. *Annu. Rev. Physiol.* 2007;69:291–316.

[13] Prosser RA. Serotonin phase-shifts the mouse suprachiasmatic circadian clock in vitro. *Brain Res.* 2003 Mar 14;966(1):110–5.

[14] Tomioka K,. Uryu O, Kamae Y, Umezaki Y, Yoshii T. *Peripheral circadian rhythms and their regulatory mechanism in insects and some other arthropods: a review.* 2012 Aug;729–40.

[15] Gillette MU, DeMarco SJ, Ding JM, Gallman EA, Faiman LE, Liu C, et al. The organization of the suprachiasmatic circadian pacemaker of the rat and its regulation by neurotransmitters and modulators. 1993;S53–58.

[16] Barnes S, Jacklet JW. Ionic currents of isolated retinal pacemaker neurons: projected daily phase differences and selective enhancement by a phase-shifting neurotransmitter. 1997 Jun;3075–84.

[17] Arango V, Underwood MD, Mann JJ. Serotonin brain circuits involved in major depression and suicide. Prog. Brain Res. 2002;136:443–53.

[18] Brieden T, Ujeyl M, Naber D. Psychopharmacological treatment of aggression in schizophrenic patients. *Pharmacopsychiatry.* 2002 May;35(3):83–9.

[19] Jenike MA, Rauch SL, Cummings JL, Savage CR, Goodman WK. Recent developments in neurobiology of obsessive-compulsive disorder. *J Clin Psychiatry.* 1996 Oct;57(10):492–503.

[20] Sullivan JM, Genco MC, Marlow ED, Benton JL, Beltz BS, Sandeman DC. Brain photoreceptor pathways contributing to circadian rhythmicity in crayfish. 2009 Aug;1136–68.

[21] Valdés-Fuentes M, Prieto-Sagredo J, Fanjul-Moles ML. Crayfish brain-protocerebrum and retina show serotonergic functional relationship. *Brain Res.* 2011 Oct 12;1417:36–44.

[22] Calderón-Rosete G, Flores G, Rodríguez-Sosa L. Diurnal rhythm in the levels of the serotonin 5-HT1A receptors in the crayfish eyestalk. *Synapse.* 2006 May;59(6):368–73.

[23] Aréchiga H, Bañuelos E, Frixione E, Picones A, Rodríguez-Sosa L. Modulation of crayfish retinal sensitivity by 5-hydroxytryptamine. 1990 May;123–43.

[24] Rodríguez-Sosa, Picones, Rosete, AréchigaY. Localization and release of 5-hydroxytryptamine in the crayfish eyestalk. *J. Exp. Biol.* 1997 Dec;200 (Pt 23):3067–77.

[25] Sandeman DC, Sandeman RE, De Couet HG. Extraretinal photoreceptors in the brain of the crayfish *Cherax destructor.* 1990 Jun;619–29.

[26] Strauss J, Dircksen H. Circadian clocks in crustaceans: identified neuronal and cellular systems. 2010;1040–74.

[27] Villagran-Vargas E, Rodríguez-Sosa L, Hustert R, Blicher A, Laub K, Heimburg T. Variations in interpulse interval of double action potentials during propagation in single neurons. 2013 Feb;68–78.

[28] Rodríguez-Sosa L, Calderón-Rosete G, Flores G. Circadian and ultradian rhythms in the crayfish caudal photoreceptor. 2008 Sep;643–52.

[29] Rodríguez Sosa L, Calderón-Rosete G, Anaya V, Flores G. The caudal photoreceptor in crayfish: an overview. In: Akutagawa E, Ozaki K, editors. Photoreceptors physiology, types, and abnormalities. New York: Nova Science Publishers; 2012. page 59–78.

[30] Simon TW, Edwards DH. Light-evoked walking in crayfish: behavioral and neuronal responses triggered by the caudal photoreceptor. *Journal of Comparative Physiology A*. 1990;166:745–55.

[31] Edwards DH Jr. Crayfish extraretinal photoreception. I. Behavioral and motorneuronal responses to abdominal illumination. *J. Exp. Biol.* 1984 Mar;109:291–306.

[32] Simon TW, Edwards DH. The Caudal Photoreceptor: A Multifunctional Sensory Neuron may select its Outputs by Spike Frequency. In: Wiese K, Krenz WD, Tautz J, Reichert H, Mulloney B, editors. Frontiers in Crustacean Neurobiology. Boston, Berlin: Birkhauser-Verlag; 1990.

[33] Fernández de Miguel F, Aréchiga H. Sensory inputs mediating two opposite behavioural responses to light in the crayfish Procambarus clarkii. *Journal of Experimental Biology*. 1992;164:153–69.

[34] Fernández de Miguel F, Cohen J, Zamora L, Aréchiga H. An automated system for detection and analysis of locomotor behavior in crustaceans. *Bol Estud Med Biol.* 1989 Dec;37(3-4):71–6.

[35] Aréchiga H, Fernández-Quiróz F, Fernández de Miguel F, Rodríguez-Sosa L. The circadian system of crustaceans. *Chronobiol. Int.* 1993 Feb;10(1):1–19.

[36] Rodríguez-Sosa L, Calderón-Rosete G, Flores G, Porras MG. Serotonin-caused phase shift of circadian rhythmicity in a photosensitive *neuron.* 2007 Oct;801–8.

[37] Rodríguez-Sosa L, Calderón-Rosete G, Porras Villalobos MG, Mendoza Zamora E, Anaya González V. Serotonin modulation of caudal photoreceptor in crayfish. 2006 Apr;220–30.

[38] Livingstone MS, Harris-Warrick RM, Kravitz EA. Serotonin and octopamine produce opposite postures in lobsters. *Science.* 1980 Apr 4;208(4439):76–9.

[39] Elofsson R, Laxmyr L, Rosengren E, Hansson C. Identification and quantitative measurements of biogenic amines and DOPA in the central nervous system and haemolymph of the crayfish Pacif ast acus leniusculus (crustacea). *Comparative Biochemistry and Physiology Part C: Comparative Pharmacology.* 1982 Jan;71(2):195–201.

[40] Keller R, Beyer J. Zur hyperglykämischen Wirkung von Serotonin und Augenstielextrakt beim Flusskrebs Orconectes limosus. *Zeitschrift Vergleichende Physiologie.* 1968;59(1):78–85.

[41] Fanjul-Moles ML. Biochemical and functional aspects of crustacean hyperglycemic hormone in decapod crustaceans: review and update. *Comp. Biochem. Physiol. C Toxicol. Pharmacol.* 2006 Apr;142(3-4):390–400.

[42] Fingerman M, Nagabhushanam R. Control of the release of crustacean hormones by neuroregulators. Comparative Biochemistry and Physiology Part C: *Comparative Pharmacology.* 1992 Jul;102(3):343–52.

[43] Sarojini R, Nagabhushanam R, Devi M, Fingerman M. Dopaminergic inhibition of 5-hydroxytryptamine-stimulated testicular maturation in the fiddler crab, Uca pugilator. Comparative Biochemistry and Physiology Part C: *Pharmacology, Toxicology and Endocrinology.* 1995 Jun;111(2):287–92.

[44] Sarojini R, Nagabhushanam R, Fingerman M. In vivo inhibition by dopamine of 5-hydroxytryptamine-stimulated ovarian maturation in the red swamp crayfish,Procambarus clarkii. *Experientia.* 1995 Feb;51(2):156–8.

[45] Farca Luna AJ, Heinrich R, Reischig T. The circadian biology of the marbled crayfish. *Front Biosci* (Elite Ed). 2010;2:1414–31.

[46] Aréchiga H, Rodríguez-Sosa L. Distributed circadian rhythmicity in the crustacean nervous system. In: Wise K, editor. Crustacean *Nervous System*. Berlin: Springer-Verlag; 2002. page 113–23.

[47] Sáenz F, García U, Aréchiga U. Modulation of electrical activity by 5-hydroxytryptamine in crayfish neurosecretory cells. *J. Exp. Biol.* 1997 Dec;200(Pt 23):3079–90.

[48] Alvarado-Alvarez R, Aréchiga H, García U. Serotonin activates a Ca(2+)-dependent K(+) current in identified peptidergic neurons from the crayfish. *J. Exp. Biol.* 2000 Feb;203(Pt 4):715–23.

[49] Nichols CD. 5-HT2 receptors in Drosophila are expressed in the brain and modulate aspects of circadian behaviors. *Dev Neurobiol.* 2007 May;67(6):752–63.

[50] Yuan Q, Lin F, Zheng X, Sehgal A. Serotonin modulates circadian entrainment in Drosophila. Neuron. 2005 Jul 7;47(1):115–27.

[51] Suri V, Qian Z, Hall JC, Rosbash M. Evidence that the TIM light response is relevant to light-induced phase shifts in Drosophila melanogaster. *Neuron*. 1998 Jul;21(1):225–34.

[52] Yang Z, Emerson M, Su HS, Sehgal A. Response of the timeless protein to light correlates with behavioral entrainment and suggests a nonvisual pathway for circadian photoreception. *Neuron*. 1998 Jul;21(1):215–23.

[53] Lin FJ, Song W, Meyer-Bernstein E, Naidoo N, Sehgal A. Photic signaling by cryptochrome in the Drosophila circadian system. *Mol. Cell. Biol*. 2001 Nov;21(21):7287–94.

[54] Williams JA, Sehgal A. Molecular components of the circadian system in Drosophila. *Annu. Rev. Physiol*. 2001;63:729–55.

[55] Rodriguez Moncalvo VG, Campos AR. Role of serotonergic neurons in the Drosophila larval response to light. *BMC Neurosci*. 2009;10:66.

[56] Cattaert D, Delbecque J-P, Edwards DH, Issa FA. Social interactions determine postural network sensitivity to 5-HT. *J. Neurosci*. 2010 Apr 21;30(16):5603–16.

[57] Elofsson R. 5-HT-like immunoreactivity in the central nervous system of the crayfish, Pacifastacus leniusculus. *Cell Tissue Res*. 1983;232(1):221–36.

[58] Beltz BS, Kravitz EA. Mapping of serotonin-like immunoreactivity in the lobster nervous system. *J. Neurosci*. 1983 Mar;3(3):585–602.

[59] Sandeman DC, Sandeman RE, Aitken AR. Atlas of serotonin-containing neurons in the optic lobes and brain of the crayfish, Cherax destructor. *J. Comp. Neurol*. 1988 Mar 22;269(4):465–78.

[60] Bellon-Humbert C, Van Herp F. Localization of serotonin-like immunoreactivity in the eyestalk of the prawn Palaemon serratus (Crustacea, Decapoda, Natantia). *J. Morphol*. 1988 Jun;196(3):307–20.

[61] Wiersma Ca, Ikeda K. Interneurons Commanding Swimmeret Movements In The Crayfish, Procambarus Clarki (Girard). *Comp. Biochem. Physiol*. 1964 Aug;12:509–25.

[62] Harris-Warrick RM, Kravitz EA. Cellular mechanisms for modulation of posture by octopamine and serotonin in the lobster. *J. Neurosci*. 1984 Aug;4(8):1976–93.

[63] Glusman S, Kravitz EA. The action of serotonin on excitatory nerve terminals in lobster nerve-muscle preparations. *J. Physiol*. (Lond.). 1982 Apr;325:223–41.

[64] Goy MF, Kravitz EA. Cyclic AMP only partially mediates the actions of serotonin at lobster neuromuscular junctions. *J. Neurosci.* 1989 Jan;9(1):369–79.

[65] Dixon D, Atwood HL. Phosphatidylinositol system's role in serotonin-induced facilitation at the crayfish neuromuscular junction. *Journal of Neurophysiology*. 1989;62(1):239–46.

[66] Huber R, Orzeszyna M, Pokorny N, Kravitz EA. Biogenic amines and aggression: experimental approaches in crustaceans. *Brain Behav. Evol.* 1997;50 Suppl 1:60–8.

[67] Glanzman DL, Krasne FB. 5,7-Dihydroxytryptamine lesions of crayfish serotonin-containing neurons: effect on the lateral giant escape reaction. *J. Neurosci.* 1986 Jun;6(6):1560–9.

[68] Edwards DH, Heitler WJ, Krasne FB. Fifty years of a command neuron: the neurobiology of escape behavior in the crayfish. *Trends Neurosci.* 1999 Apr;22(4):153–61.

[69] Yeh SR, Fricke RA, Edwards DH. The effect of social experience on serotonergic modulation of the escape circuit of crayfish. *Science*. 1996 Jan 19;271(5247):366–9.

[70] Yeh SR, Musolf BE, Edwards DH. Neuronal adaptations to changes in the social dominance status of crayfish. *J. Neurosci.* 1997 Jan 15;17(2):697–708.

[71] Edwards DH, Yeh S-R, Musolf BE, Antonsen BL, Krasne FB. Metamodulation of the crayfish escape circuit. *Brain Behav. Evol.* 2002;60(6):360–9.

[72] Teshiba T, Shamsian A, Yashar B, Yeh SR, Edwards DH, Krasne FB. Dual and opposing modulatory effects of serotonin on crayfish lateral giant escape command neurons. *J. Neurosci.* 2001 Jun 15;21(12): 4523–9.

[73] Krasne FB, Edwards DH. Modulation of the crayfish escape reflex--physiology and neuroethology. Integr. Comp. Biol. 2002 Aug;42(4):705–15.

[74] Hörner M, Weiger WA, Edwards DH, Kravitz EA. Excitation of identified serotonergic neurons by escape command neurons in lobsters. *J. Exp. Biol.* 1997 Jul;200(Pt 14):2017–33.

[75] Chen S, Lee AY, Bowens NM, Huber R, Kravitz EA. Fighting fruit flies: a model system for the study of aggression. *Proc. Natl. Acad. Sci. U.S.A*. 2002 Apr 16;99(8):5664–8.

[76] Yurkovic A, Wang O, Basu AC, Kravitz EA. Learning and memory associated with aggression in Drosophila melanogaster. *Proc. Natl. Acad. Sci. USA.* 2006 Nov 14;103(46):17519–24.

[77] De Boer SF, Caramaschi D, Natarajan D, Koolhaas JM. The vicious cycle towards violence: focus on the negative feedback mechanisms of brain serotonin neurotransmission. *Front Behav Neurosci.* 2009;3:52.

[78] Miczek KA, De Almeida RMM, Kravitz EA, Rissman EF, De Boer SF, Raine A. Neurobiology of escalated aggression and violence. *J. Neurosci.* 2007 Oct 31;27(44):11803–6.

[79] Dierick HA, Greenspan RJ. Serotonin and neuropeptide F have opposite modulatory effects on fly aggression. *Nat. Genet.* 2007 May;39(5):678–82.

[80] Alekseyenko OV, Lee C, Kravitz EA. Targeted manipulation of serotonergic neurotransmission affects the escalation of aggression in adult male Drosophila melanogaster. *PLoS ONE.* 2010;5(5):e10806.

[81] Lee G, Hall JC. Abnormalities of male-specific FRU protein and serotonin expression in the CNS of fruitless mutants in Drosophila. *J. Neurosci.* 2001 Jan 15;21(2):513–26.

[82] Fratta W, Biggio G, Gessa GL. Homosexual mounting behavior induced in male rats and rabbits by a tryptophan-free diet. *Life Sci.* 1977 Aug 1;21(3):379–83.

[83] Gessa GL, Tagliamonte A. Possible role of brain serotonin and dopamine in controlling male sexual behavior. *Adv. Biochem. Psychopharmacol.* 1974;11(0):217–28.

[84] Friesen WO, Poon M, Stent GS. An oscillatory neuronal circuit generating a locomotory rhythm. *Proc. Natl. Acad. Sci. U.S.A.* 1976 Oct;73(10):3734–8.

[85] Nusbaum MP, Kristan WB Jr. Swim initiation in the leech by serotonin-containing interneurones, cells 21 and 61. *J. Exp. Biol.* 1986 May;122:277–302.

[86] Szczupak L, Kristan WB Jr. Widespread mechanosensory activation of the serotonergic system of the medicinal leech. *J. Neurophysiol.* 1995 Dec;74(6):2614–24.

[87] Velázquez-Ulloa N, Blackshaw SE, Szczupak L, Trueta C, García E, De-Miguel FF. Convergence of mechanosensory inputs onto neuromodulatory serotonergic neurons in the leech. *J. Neurobiol.* 2003 Mar;54(4):604–17.

[88] Nusbaum MP. Synaptic basis of swim initiation in the leech. III. Synaptic effects of serotonin-containing interneurones (cells 21 and 61)

on swim CPG neurones (cells 18 and 208). *J. Exp. Biol.* 1986 May;122:303–21.
[89] Angstadt JD, Friesen WO. Modulation of swimming behavior in the medicinal leech. I. Effects of serotonin on the electrical properties of swim-gating cell 204. *J. Comp. Physiol. A.* 1993 Mar;172(2):223–34.
[90] Angstadt JD, Friesen WO. Modulation of swimming behavior in the medicinal leech. II. Ionic conductances underlying serotonergic modulation of swim-gating cell 204. *J. Comp. Physiol. A.* 1993 Mar;172(2):235–48.
[91] Mangan PS, Cometa AK, Friesen WO. Modulation of swimming behavior in the medicinal leech. IV. Serotonin-induced alteration of synaptic interactions between neurons of the swim circuit. *J. Comp. Physiol. A.* 1994 Dec;175(6):723–36.
[92] Lent CM, Dickinson MH. Retzius cells retain functional membrane properties following "ablation" by the neurotoxin 5,7-DHT. *Brain Res.* 1984 May 21;300(1):167–71.
[93] Mar A, Drapeau P. Modulation of conduction block in leech mechanosensory neurons. *J. Neurosci.* 1996 Jul 15;16(14):4335–43.
[94] Wilson RJ, Kristan WB Jr, Kleinhaus AL. An increase in activity of serotonergic Retzius neurones may not be necessary for the consummatory phase of feeding in the leech Hirudo medicinalis. *J. Exp. Biol.* 1996 Jun;199(Pt 6):1405–14.
[95] Sawada M, Coggeshall RE. A central inhibitory action of 5-hydroxytryptamine in the leech. *J. Neurobiol.* 1976 Nov;7(6):477–82.
[96] Sawada M, Coggeshall RE. Ionic mechanism of 5-hydroxytryptamine induced hyperpolarization and inhibitory junctional potential in body wall muscle cells of Hirudo medicinalis. *J. Neurobiol.* 1976 Jan;7(1):63–73.
[97] Lent CM. Retzius cells: neuroeffectors controlling mucus release by the leech. *Science.* 1973 Feb 16;179(4074):693–6.
[98] Groome JR, Clark M, Lent CM. The behavioural state of satiation in the leech is regulated by body distension and mimicked by serotonin depletion. *J. Exp. Biol.* 1993 Sep;182:265–70.
[99] Neckameyer WS. A trophic role for serotonin in the development of a simple feeding circuit. *Dev. Neurosci.* 2010 Aug;32(3):217–37.
[100] Ehrlich JS, Boulis NM, Karrer T, Sahley CL. Differential effects of serotonin depletion on sensitization and dishabituation in the leech, Hirudo medicinalis. *J. Neurobiol.* 1992 Apr;23(3):270–9.

[101] Sahley CL. Serotonin depletion impairs but does not eliminate classical conditioning in the leech Hirudo medicinalis. *Behav. Neurosci.* 1994 Dec;108(6):1043–52.

[102] Burrell BD, Sahley CL, Muller KJ. Non-associative learning and serotonin induce similar bi-directional changes in excitability of a neuron critical for learning in the medicinal leech. *J. Neurosci.* 2001 Feb 15;21(4):1401–12.

[103] Burrell BD, Sahley CL, Muller KJ. Differential effects of serotonin enhance activity of an electrically coupled neural network. *J. Neurophysiol.* 2002 Jun;87(6):2889–95.

[104] Quinn WG, Harris WA, Benzer S. Conditioned behavior in Drosophila melanogaster. *Proc. Natl. Acad. Sci. U.S.A.* 1974 Mar;71(3):708–12.

[105] Tempel BL, Bonini N, Dawson DR, Quinn WG. Reward learning in normal and mutant Drosophila. *Proc. Natl. Acad. Sci. U.S.A.* 1983 Mar;80(5):1482–6.

[106] Johnson O, Becnel J, Nichols CD. Serotonin receptor activity is necessary for olfactory learning and memory in Drosophila melanogaster. *Neuroscience.* 2011 Sep 29;192:372–81.

[107] HAGIWARA S, MORITA H. Electrotonic transmission between two nerve cells in leech ganglion. *J. Neurophysiol.* 1962 Nov;25:721–31.

[108] Eckert R. Electrical Interaction of Paired Ganglion Cells in the Leech. *J. Gen. Physiol.* 1963 Jan 1;46(3):573–87.

[109] García-Pérez E, Vargas-Caballero M, Velazquez-Ulloa N, Minzoni A, De-Miguel FF. Synaptic integration in electrically coupled neurons. *Biophys. J.* 2004 Jan;86(1 Pt 1):646–55.

[110] De-Miguel FF, Vargas-Caballero M, García-Pérez E. Spread of synaptic potentials through electrical synapses in Retzius neurones of the leech. *J. Exp. Biol.* 2001 Oct;204(Pt 19):3241–50.

[111] Vazquez Y, Mendez B, Trueta C, De-Miguel FF. Summation of excitatory postsynaptic potentials in electrically-coupled neurones. *Neuroscience.* 2009 Sep 29;163(1):202–12.

[112] Lent CM, Frazer BM. Connectivity of the monoamine-containing neurones in central nervous system of leech. *Nature.* 1977 Apr 28;266(5605):844–7.

[113] Nicholls JG, Kuffler DP. Quantal release of serotonin from presynaptic nerve terminals. *Neurochem. Int.* 1990;17(2):157–63.

[114] Fernández-de-Miguel F, Drapeau P. Synapse formation and function: insights from identified leech neurons in culture. *J. Neurobiol.* 1995 Jul;27(3):367–79.

[115] Henderson LP. The role of 5-hydroxytryptamine as a transmitter between identified leech neurones in culture. *J. Physiol.* (Lond.). 1983 Jun;339:309–24.

[116] Henderson LP, Kuffler DP, Nicholls J, Zhang R. Structural and functional analysis of synaptic transmission between identified leech neurones in culture. *J. Physiol.* (Lond.). 1983 Jul;340:347–58.

[117] Liu Y, Nicholls J. Steps in the development of chemical and electrical synapses by pairs of identified leech neurons in culture. *Proc. R. Soc. Lond., B, Biol. Sci.* 1989 Apr 22;236(1284):253–68.

[118] Dietzel ID, Drapeau P, Nicholls JG. Voltage dependence of 5-hydroxytryptamine release at a synapse between identified leech neurones in culture. *J. Physiol.* (Lond.). 1986 Mar;372:191–205.

[119] Kuffler DP, Nicholls J, Drapeau P. Transmitter localization and vesicle turnover at a serotoninergic synapse between identified leech neurons in culture. *J. Comp. Neurol.* 1987 Feb 22;256(4):516–26.

[120] Bruns D, Jahn R. Real-time measurement of transmitter release from single synaptic vesicles. *Nature*. 1995 Sep 7;377(6544):62–5.

[121] Cercós MG, De-Miguel FF, Trueta C. Real-time measurements of synaptic autoinhibition produced by serotonin release in cultured leech neurons. *J. Neurophysiol.* 2009 Aug;102(2):1075–85.

[122] Agnati LF, Genedani S, Leo G, Forni A, Woods AS, Filaferro M, et al. Abeta peptides as one of the crucial volume transmission signals in the trophic units and their interactions with homocysteine. Physiological implications and relevance for Alzheimer's disease. *J Neural Transm.* 2007 Jan;114(1):21–31.

[123] Fuxe K, Borroto-Escuela DO, Romero-Fernandez W, Diaz-Cabiale Z, Rivera A, Ferraro L, et al. Extrasynaptic neurotransmission in the modulation of brain function. Focus on the striatal neuronal-glial networks. 2012;136.

[124] Trueta C, De-Miguel FF. Extrasynaptic exocytosis and its mechanisms: a source of molecules mediating volume transmission in the nervous system. Frontiers in Physiology [Internet]. 2012 [cited 2012 Dec 20];3. Available from: http://www.frontiersin.org/Membrane_Physiology_and_Biophysics/10.3389/fphys.2012.00319/abstract

[125] Trueta C, Méndez B, De-Miguel FF. Somatic exocytosis of serotonin mediated by L-type calcium channels in cultured leech neurones. *J. Physiol.* (Lond.). 2003 Mar 1;547(Pt 2):405–16.

[126] Coggeshall RE. Autoradiographic and chemical localization of 5-hydroxytryptamine in identified neurons in the leech. *Anat. Rec.* 1972 Mar;172(3):489–98.

[127] Trueta C, Sánchez-Armass S, Morales MA, De-Miguel FF. Calcium-induced calcium release contributes to somatic secretion of serotonin in leech Retzius neurons. *J. Neurobiol.* 2004 Dec;61(3):309–16.

[128] Trueta C, Kuffler DP, De-Miguel FF. Cycling of dense core vesicles involved in somatic exocytosis of serotonin by leech neurons. *Front Physiol.* 2012;3:175.

[129] Patel JC, Witkovsky P, Avshalumov MV, Rice ME. Mobilization of calcium from intracellular stores facilitates somatodendritic dopamine release. *J. Neurosci.* 2009 May 20;29(20):6568–79.

[130] Ludwig M, Sabatier N, Bull PM, Landgraf R, Dayanithi G, Leng G. Intracellular calcium stores regulate activity-dependent neuropeptide release from dendrites. *Nature.* 2002 Jul 4;418(6893):85–9.

[131] De-Miguel FF, Santamaría-Holek I, Noguez P, Bustos C, Hernández-Lemus E, Rubí JM. Biophysics of active vesicle transport, an intermediate step that couples excitation and exocytosis of serotonin in the neuronal soma. *PLoS ONE.* 2012;7(10):e45454.

[132] Kaushalya SK, Desai R, Arumugam S, Ghosh H, Balaji J, Maiti S. Three-photon microscopy shows that somatic release can be a quantitatively significant component of serotonergic neurotransmission in the mammalian brain. *J. Neurosci. Res.* 2008 Nov 15;86(15):3469–80.

[133] Maiti S. Measuring Serotonin Distribution in Live Cells with Three-Photon Excitation. *Science.* 1997 Jan 24;275(5299):530–2.

[134] Nicholls JG, Baylor DA. Specific modalities and receptive fields of sensory neurons in CNS of the leech. *J. Neurophysiol.* 1968 Sep;31(5):740–56.

[135] Yau KW. Receptive fields, geometry and conduction block of sensory neurones in the central nervous system of the leech. *J. Physiol.* (Lond.). 1976 Dec;263(3):513–38.

[136] Yau KW. Physiological properties and receptive fields of mechanosensory neurones in the head ganglion of the leech: comparison with homologous cells in segmental ganglia. *J. Physiol.* (Lond.). 1976 Dec;263(3):489–512.

[137] Blackshaw SE. Morphology and distribution of touch cell terminals in the skin of the leech. *J. Physiol.* (Lond.). 1981 Nov;320:219–28.

In: New Developments in Sensory Processing … ISBN: 978-1-62808-395-8
Editors: M. Voisin and R. Brunel

Chapter III

Sensory Modalities Processing in Sensorimotor Tasks Depends on State and Trait Anxiety

Jean-Philippe Hainaut*[*] *and Benoît Bolmont
Université de Lorraine,
UFR Sciences Fondamentales et Appliquées,
Campus Bridoux, Metz, France
Emotions-Actions, LCOMS (EA 7306)

Abstract

In the daily life activities, the ability to divide attention between different sensory modalities is essential since we have to cope with tasks that require multisensory integration. An individual has to select the relevant information, and sometimes shifts their attention from one sensory modality to another when one sensory modality is inadequate, absent, or salient. These sensory reweighting mechanisms are needed to maintain an efficient performance. Previous works have suggested that the highly complex attentional system that is needed in such situations could be influenced by affective states, such as anxiety. Indeed, it has

[*] Corresponding author: Tel/ Fax: 33 387 378 624/33 387 378 603, E-mail: jean-philippe.hainaut@univ-lorraine.fr.

been shown that an anxious state influences the efficient functioning of the attentional system. This can result in differences in the processing of sensory modalities.

Moreover, recent works have highlighted that the allocation of attentional resources towards sensory modalities can be modulated by personality traits, such as trait-anxiety. This review focuses on the interdependence between anxiety state, anxiety trait and sensory modalities processing during tasks that require multisensory integration (*e.g.* sensorimotor tasks, balance control, posture...)

Introduction

In everyday life, most of our activities require us to simultaneously integrate and process various sensory inputs. This mechanism of multisensory integration is essential, since it allows us to perceive a world of coherent perceptual entities [56]. In order to achieve our behavioural goals, we must appropriately select and integrate the sensory inputs that are the most relevant to successful behaviours.

In this context, the interactions between multisensory processing and attentional selection systems are fundamental for an effective functioning [96]. Since the 1960s, selective attention research was largely concerned in studying each sensory modality in strict isolation. The recent decade has seen an important increase of research activity in crossmodal attention and multisensory processing (*e.g.*, [3, 15, 51, 74, 96]).

In normal conditions, the various sensory systems provide complementary and redundant information about the environment. Individual has to select relevant information, and sometimes shifts his attention from one sensory modality to another [21, 90]. Among attentional mechanisms that lead individual to shift his attention, it is conventional to distinguish between exogenous mechanisms of attention (driven by salient stimuli), versus endogenous mechanisms [81], that allow to select continuously from among incoming sensory stimuli, the relevant stimuli that will benefit from faster and heightened processing [3]. These sensory reweighting mechanisms are needed to maintain an efficient functioning.

These attentional mechanisms are central in many sensorimotor tasks. For example, numerous studies have investigated the weight of sensory inputs involved in balance control [20, 23-25, 44, 45, 70, 77, 78]. Because information from the various sensory inputs involved in balance control is not always available or accurate, the weight assigned to each modality may need to be adjusted -or reweighted-, in order to maintain stance in a wide variety of environmental conditions [77, 78].

This perceptual inhibition (*i.e.* the ability to centrally suppress the inaccurate sensory input) is highly linked to the ability to shift attention between sensory modalities [84, 85].

As well as other cognitive mechanisms, these multisensory mechanisms could depend on psychological states [6, 10, 43, 71, 101, 107]. Among the psychological states, anxiety is a complex phenomenon that influences various aspects of cognition including the attentional systems. Indeed, numerous studies have highlighted relationships between anxiety and attention (for a review, see [29]), especially for the inhibition and the shifting functions [62]. Since these two functions are central in multisensory integration, we hypothesized that anxiety could greatly influence the allocation of attentional resources toward one or other sensory modality.

Anxiety, which could be considered with low moods in a single construct [8], is a multidimensional phenomenon that can be divided into state and trait dimensions. As stated by Spielberger [91], while state anxiety is transitory, trait anxiety may be conceived as latent dispositions to perceive and to behave in a specific manner.

According to Spielberger, personality traits will be manifested in behaviour particularly if they are triggered by appropriate stimulus [91]. In line with previous studies that have demonstrated that groups of different trait anxiety level will adopt different behaviours when exposed to a stressor [68, 105], we conducted some experiments with subjects possessing different trait anxiety level, in order to determine the effect of state anxiety on multisensory integration underlying various sensorimotor tasks.

The behavioural consequences and mechanisms underlying the effects of trait and state anxiety on multisensory integration provide the focus of this article. The relationship between anxiety and sensorimotor performances involving different sensory modalities is discussed.

In a first section, we examine the effects of anxiety on attentional mechanisms involved in response time performances of different sensory modalities. In a second section, we focus on the role of trait anxiety on balance control in control and anxiogenic condition.

State Anxiety, Trait Anxiety and Attentional Mechanisms Underlying Response Time Performance in Different Sensory Modalities

Anxiety and Response Time Performances

The relationship between anxiety and sensorimotor performances has been widely investigated and a complex and contrasted relationship has been highlighted [7, 10, 11, 28, 47, 71, 101]. On the basis of the inverted-U theory [109] suggesting that sensorimotor performance is maximized at intermediate levels of anxiety [59, 82], Oxendine [73] has suggested that simple motor task requiring rapidity or strength would be maximized at high level of state-anxiety, whereas complex motor skills would be altered.

As far as response time tasks are concerned, it has been suggested that in tasks with high information content such as complex auditory and visual choice response time tasks, the state anxiety may produce a slowing down of the information processing and motor strategies [31, 47]. These deficits in performance of complex tasks may be due to alterations in higher brain function [1]. Alternatively, state anxiety would lead to better performances in visual stimulus-response tasks such as two-choice response tasks, which depend greatly on automatic abilities, by recruiting motivational capacities [7]. In fact, high state anxiety provokes attentional selectivity [4, 76], which in turn would favor the information processing in binary visual choice reaction time task [7]. Besides, previous data have provided insight into the interactions between trait anxiety and response time performances [9]. Trait anxiety favours the processing of information of stimulus-response tasks, but not that of controlled tasks that required strategic processes [9]. According to Spielberger, personality traits, that may be conceived as latent dispositions to perceive and to behave in a specific manner, will be manifested in behaviour particularly if they are triggered by appropriate stimulus [91]. In this way, a study comparing groups of different trait anxiety levels during normal and anxiogenic conditions has indicated that arousal and the impairment of the accuracy in performing a motor task increase with trait anxiety, particularly when the intensity of the stressor elevates [68]. Nevertheless, to the best of our knowledge, there is only very limited research that have investigated and discussed the interrelation between state anxiety, trait anxiety and response time tasks.

Thus, we have conducted two studies in order to assess in subjects possessing contrasted levels of trait anxiety whether moderate state anxiety influences reaction or movement time in response time tasks of different sensory modalities. While the first study concentrated on simple response time tasks in visual and auditory modalities (each assessed separately) [35], the second study used a bimodal choice response time task (visual or auditory) [36] that could be closer to daily life situations in which attention must be coordinated between auditory and visual modalities.

The induction of moderate state anxiety was made by the Stroop Color-Word Test with interferences, combined with video recording. This association has been shown to induce transitory psychological and physiological modifications [34, 88, 97, 98, 104], whatever the subjects' trait-anxiety [55, 105]. The Stroop Color-Word Test with interferences was performed before the response time tasks. Indeed previous studies have shown a persistence of the physiological arousal after that the stressful Stroop task has been completed [104, 105].

In these two studies, the subjects whose trait anxiety scores (STAI [92]) were at the extremes of the distribution were assigned in two separate groups. According to the Spielberger's classification [92], individuals were shifted as subjects with a very low trait anxiety (VLTA) level or an intermediate trait anxiety (ITA) level. Response time tasks were performed in control and anxiogenic conditions. Response times were distinguished for two components: (i) reaction time, *i.e.* the time required to evaluate the stimulus, and to select and trigger the response, and (ii) movement time, *i.e.* the time required to execute the motor response [46].

Unimodal Sensory Processing: Simple Response Time Tasks in Visual and Auditory Modalities

The main finding of this study focusing on simple reaction time in visual and auditory modalities each assessed separately, is that moderate state-anxiety differently improves performance in individuals possessing intermediate trait anxiety or very low trait anxiety level. While VLTA subjects improved performance in auditory response time, ITA subjects improved performance in visual response time (see figure 1). In both groups, improvement of performance in response time occurred at the reaction time and movement time level.

Reaction time, movement time and state anxiety

As far as reaction time is concerned, attentional mechanisms, that can strongly modulate sensory processing [22], could play a central role in the improvement of performances during the anxiogenic condition. Indeed, several studies have shown the relationship between anxiety and attention [4, 7, 28, 29, 60, 76, 87]. Anxiogenic condition may increase arousal [40, 69, 105], which could improve attention and lead subjects to attend actively to any incoming signal [19].

In line with a previous study [7], the present results suggest that state anxiety could have improved reaction time performances by modulating attentional mechanisms involved in sensory processing [22]. As far as movement time is concerned, our results suggest that the anxiogenic condition has a positive effect on the movement time as well as the reaction time. In addition to the effects on reaction time, attention could contribute to improve the movement time performances. It has been demonstrated that the increases in attention and arousal could lead to an elevated muscular tension [93, 102, 103]. Moreover state anxiety is also associated with increase in muscular tension [38, 39, 57] even when a moderate state anxiety is induced by the Stroop Color-Word Test with interferences [97, 104, 105]. In line with studies showing that an optimal muscular tension tends to shorten response time [27, 49, 93], our results suggest that a high arousal level due to the anxiogenic condition could have improved the muscular tension [39, 105] leading to better movement time performances.

Overall, our results in anxiogenic condition suggest that the improvement in response time performance could have a central origin (modulation of the sensory processing), and a peripheral origin (increase in muscular tension). However, in the relationship between response time and state anxiety, our results have also indicated some differences in visual and auditory tasks according to the level of trait anxiety.

State Anxiety, Trait Anxiety and Unimodal Sensory Processing

As stated above, ITA, but not VLTA, subjects improved performance in visual response time. This agrees with previous data [99] that have demonstrated in a non-anxiogenic condition that subjects with intermediate level of trait anxiety show a greater visual dependence than subjects with a very low trait anxiety level while performing a perceptual-motor task. In addition, given that behaviours associated with trait anxiety are manifest during anxiogenic conditions [91], it is likely that improved performance in visual response time produced by moderate state anxiety in ITA subjects may have resulted from a greater arousal [69] and attention for visual stimuli. Alternatively, VLTA, but not ITA, subjects improved performance in auditory response time.

This suggests that VLTA subjects exposed to anxiogenic conditions may focus on auditory stimuli rather than on visual stimuli. It should be noted that in normal condition, auditory stimuli, that have an automatic alerting effect, are processed less actively than visual stimuli [18, 50, 80].

Conversely, in anxiogenic condition, the general arousal induced by the state anxiety could have led VLTA subjects to develop more attentional resources and to process consciously auditory stimuli as well as visual stimuli. In this context, the improvement of auditory response time task in VLTA group could reflect some changes in attentional processing of this modality.

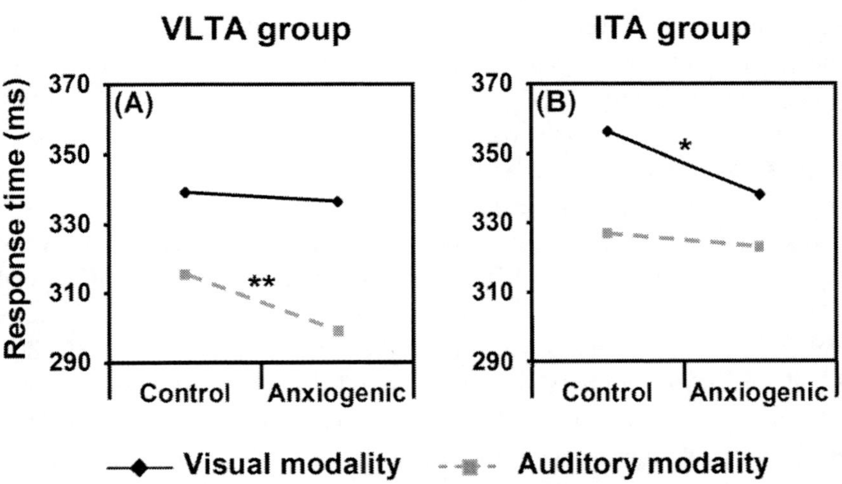

* $p < 0.05$, ** $p < 0.01$ Significant differences between the control condition and the anxiogenic condition.

Figure 1. Response time performances (mean in ms), *i.e.* the sum of the reaction time and the movement time, in the control condition and the anxiogenic condition for the very low trait anxiety subjects (VLTA; A) and the intermediate trait anxiety subjects (ITA; B).

Multisensory Processing: Bimodal Choice Response Time Tasks in Visual or Auditory Modalities

Although each sensory modality can be considered in strict isolation, in many situations people have to process simultaneously information from various modalities. Therefore, our second study [36] aimed to determine thoroughly by which anxiety may influence reaction time and/or movement time in a crossmodal task. The subjects' performances in response time were assessed using a bimodal choice task providing either visual or auditory modality. The subjects were required to press as quickly as possible the response button corresponding to the stimulus.

Relevant Modality According to Trait Anxiety

Contrary to our first study, statistical analysis did not show any main effect of condition on both reaction and movement times. Although no significant effect was detected on reaction time, we found that moderate state anxiety improves movement time performances in visual modality in ITA subjects, as well as movement time performances in auditory modality in VLTA subjects (see figure 2).

This confirms the results of our precedent study cited above [35], on the relationships between state anxiety, trait anxiety and response time to visual and auditory stimuli given separately [5]. Taken together, these two studies suggest that, whether or not a decision-making is required, individuals exposed to anxiogenic conditions could favour the information processing of a specific "relevant" modality, either visual or auditory, depending on their trait anxiety level. As such, ITA subjects would focus their attention towards visual stimuli, and VLTA subjects towards auditory stimuli. It should be noted that in similar tasks where attention is divided between visual and auditory modalities, Driver has shown that subjects mainly attend to one modality versus another [21]. Given our findings, we suggest that this may be related to the subjects' trait anxiety level.

In addition according to the work of Spielberger [91] who has suggested that a specific behaviour would be amplified if it is triggered by an appropriate stimulus, we hypothesise that these effects could be intensified in anxiogenic conditions.

Paradoxically, in anxiogenic condition, improvement of performance in both groups occurred at movement time, but not at reaction time in the preferential modality (See Figure 2). According to Driver [21], some crossmodal selection could apparently arise before diffusion of the stimulus. In this way, we suggest that subjects would prepare themselves to react to the expected modality, and thereby could start early motor program for this modality. Thus, this could contribute to facilitate movement execution when the expected modality corresponds to the presented modality, but not when the presented modality differs from the expected modality. However, despite the subjects' oriented attention towards a specific modality, we found no improvement in reaction time. We suggest that this could be due to a stage of verification that the presented modality really corresponds to the expected modality.

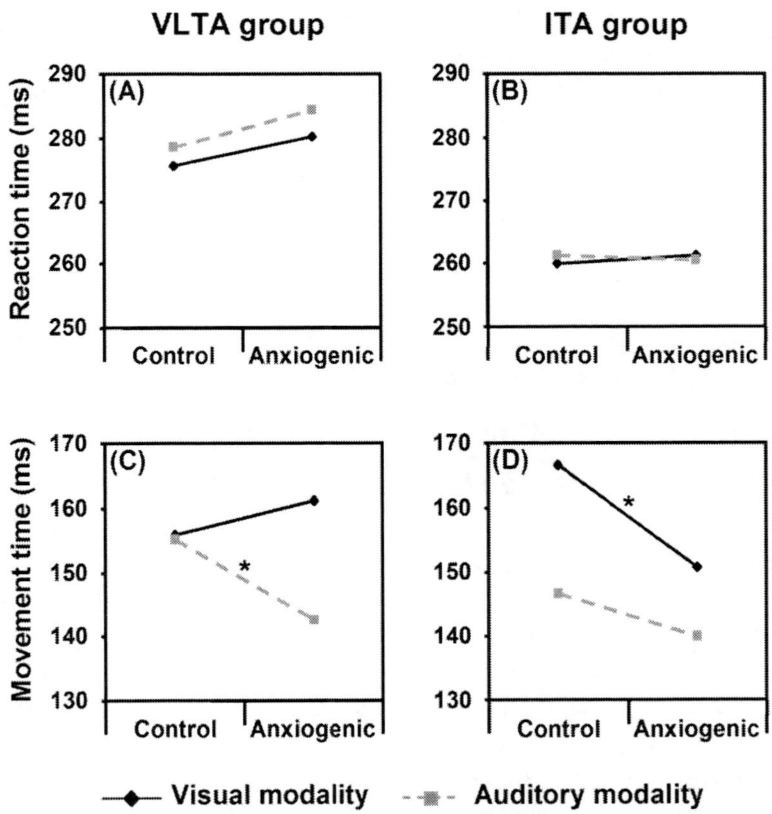

* $p < 0.05$, significant difference between the control condition and the anxiogenic condition.

Figure 2. Differences of reaction time and movement time performances (mean in ms) between the control condition and the anxiogenic condition for the very low trait anxiety group (VLTA; A, C) and the intermediate trait anxiety group (ITA; B, D).

Conclusion

In conclusion, our results show that subjects exposed to an anxiogenic condition would allocate attentional resources towards a specific relevant modality, either visual or auditory, depending on their trait anxiety level. In simple reaction time tasks, when modalities are assessed separately, both reaction and movement time performances could be improved in an anxiogenic condition by more active stimulus processing and enhanced muscular tension [35].

In bimodal choice response time task, such attentional resources would influence movement time, but not reaction time [36]. Further experiments are needed to determine whether our findings may occur in subjects with higher state anxiety levels. However, according to the inverted-U hypothesis [59, 109] and given the human limits of adaptation, it is possible to hypothesize that in contrast with our findings, performance may decline in conditions eliciting high levels of state anxiety. Overall, these results highlight personality and individual differences as regards to the human/environment interaction.

State Anxiety, Trait Anxiety and Multisensory Integration in Balance Control

Balance Control and Multisensory Integration

Although response time tasks are very interesting models to understand the attentional mechanisms when subjects have to direct attention toward one or other sensory modality, we devised another experiment [37] in order to study relative weighting of different sensory information directly related to balance control, *i.e.*, when subjects have to perform balance tasks.

The human balance control system, that depends on the quality and the integrity of the motor effectors, results from the integration at the central nervous system level of vestibular, somatosensory, and visual inputs [75]. Each sensory system plays a major role in the control of posture. This has been well established through the clinical observation of individuals deprived of sensory inputs from one or more of the three systems. A lack of input, or a lack of accurate input, generally decreases postural control performance [106]. All these sensorial afferences allow organisation of motor reflex responses, such as the vestibulo-ocular and the vestibulo-spinal reflexes, and the occurrence of adaptive and compensatory ocular and body movements [67]. Among these sensory systems, the importance of vision is well known in postural stabilization, especially for maintaining the balance precision as measured by the Sway Area (SA). Although the SA parameter with eyes open depends on distance to target fixation (SA decreased when the distance shortened [48, 54]), classically, the SA parameter is larger with eyes closed (EC) than eyes open (EO). Indeed, the lack of visual input decreases postural performances. When their eyes are closed, subjects would compensate the lack of visual input by an ankle stiffness strategy, in order to enhance the somatosensory perception [30]. This sensory reweighting strategy could help subjects to maintain precision of balance when vision is absent.

Balance Control and Anxiety

Moreover, studies have shown that balance performance could depend on psychological states such as state anxiety [2, 10, 17, 71, 101]. The fact that patients suffering from different pathological anxiety disorders are frequently confronted with postural instability has been highlighted [33, 89]. This association between anxiety and balance disorders is not restricted to clinical samples and even occurs in healthy subjects [52, 108]. State anxiety in healthy subjects could be considered to be a transitory state of either apprehension or arousal, resulting from the anticipation of an actual or non-actual threatening event respectively. It has been shown that state anxiety caused by actual threat of a fall could lead a healthy subject to develop a postural stiffness strategy in order to avoid this fall; this strategy is characterized by a smaller amplitude and higher frequency in CFP displacements [2, 14, 16, 17]. In contrast, few studies have investigated the effect of state anxiety not directed toward potential postural threat [10, 71, 101] on static balance performance. When vision is available, state anxiety could influence interactions of visual input with vestibular and somatosensory inputs [71, 101]. Another study on this issue, one which used a computer-controlled movable platform, has also shown that state anxiety and adverse mood changes could affect the subjects' ability to use inputs from the vestibular system in order to maintain balance [10]. These studies suggest that state anxiety could influence one or more sensory inputs underlying static balance control. In line with our precedent studies cited above [35, 36] suggesting that individuals possessing different trait anxiety levels preferentially process some sensory modalities in an anxiogenic condition, it could be helpful to take into account the state anxiety and the trait anxiety, in assessing the individual sensory strategies used to maintain static balance. Thus, in agreement with these previous results and with Spielberger's concept [91], we hypothesized that the weight of sensory inputs involved in balance control could be differently affected by state anxiety, depending on trait anxiety level.

Consequently, we have conducted a study in order to determine the influence of state anxiety on static balance performance with EO and EC in two groups of healthy human subjects with contrasted trait anxiety levels (named VLTA and ITA groups, according to the Spielberger's classification [92]). As well as in our previous studies, subjects performed the tasks in control condition and in anxiogenic condition induced by the Stroop Color-Word Test with interference. We used posturographic parameters, such as Sway Area (SA) and Sway Path Length (SPL), as revealed by Centre of Foot Pressure (CFP) displacements in order to investigate balance control [100]. As far as static posturography is concerned, low values for both SPL (distance in mm covered by the CFP displacement) and SA parameters (surface in mm^2 covered by the CFP displacement) reflect efficient balance control. While the SPL parameter is closely linked to sway velocity, the SA parameter is considered to be the most representative parameter regarding precision in balance [26].

Trait Anxiety and Visual Dependence

Our results have demonstrated that a moderate state anxiety increases the SA and SPL balance parameters (i.e., larger and faster body sways) for both VLTA and ITA groups when visual input is available. In contrast, with EC, the anxiogenic condition affects balance performance differently according to the level of trait anxiety.

These results should be analysed considering the specificities according to the level of trait anxiety to maintain balance in the control condition. Indeed, our results in the control condition (see figure 3) highlight some specificities in the way balance is regulated when vision is absent. Comparisons between eyes-open and eyes-closed performances have shown that SPL increases in both groups, while SA increases in the ITA group only. In agreement with previous studies, VLTA subjects tended to compensate the lack of visual input with an ankle stiffness strategy [30]. This somatosensory regulation of balance, which increases the SPL parameter, allows the SA parameter to be maintained when the eyes are closed [54, 72]. In contrast, in this condition, the ITA subjects do not seem to use a strategy like this to compensate for the absence of visual input. These discrepancies could be explained by a visual dependence phenomenon. Studies have shown that anxious patients were visually dependent in balance performance [43, 107]. Moreover, it has been also suggested that ITA subjects have a greater visual dependence than VLTA subjects while performing a perceptual-motor task [99]. In agreement with this concept of visual dependence and our own results in the control condition showing a significant SA increase in eyes-closed compared to eyes-open performances, we suggest that the lack of visual input could disrupt the precision of static balance control in ITA subjects more than in VLTA subjects. Given this potential visual dependence, we have hypothesized that the ITA subjects could be perturbed by the EC test even in the control condition.

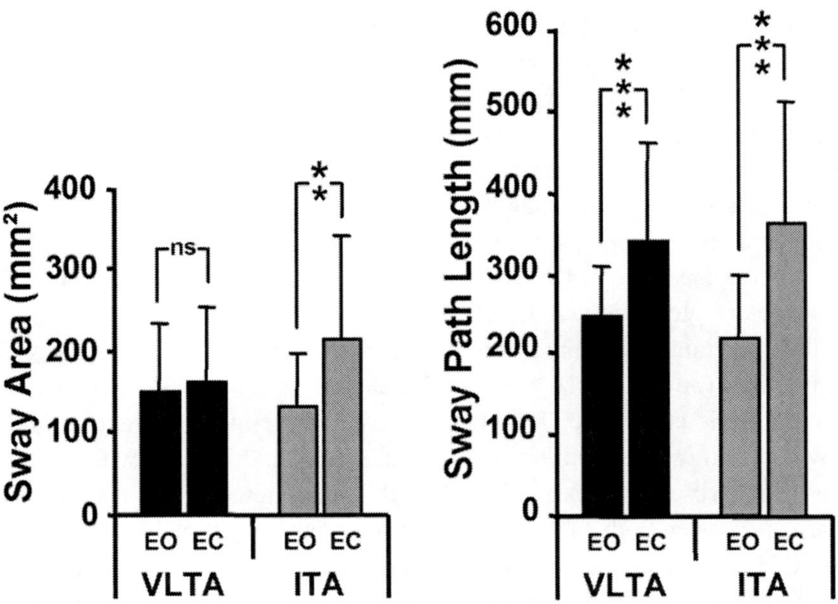

** $p < 0.01$, *** $p < 0.001$, significant difference between EO and EC tests.

Figure 3. Control condition: differences of balance performances (sway area and sway path length) between eyes-open (EO) and eyes-closed (EC) tests.

Role of Trait Anxiety in Anxiogenic Condition

When visual input is available, our results show that whatever the subjects' trait anxiety level, SA and the SPL parameters increase in the anxiogenic condition (see figure 4). Studies have shown that state anxiety could modify multisensory integration by modulating the processing of visual input [71, 101] and/or by exerting direct effects on vestibular functioning [5, 6, 42], provoking larger and faster body sways, which increase the SA and SPL parameters. In the present study, similar effects may have occurred. In addition, the somatosensory system could also be influenced by state anxiety. Previous studies showed that fear of falling induces anxiety and leads subjects to adopt a postural stiffness strategy (increased co-contraction of ankle joint agonist/antagonist muscle pairs) characterized by smaller amplitude and higher frequency in CFP displacements [2, 14, 16, 17]. In contrast, in the present study, state anxiety that is not related to a potential postural threat may have led subjects to adopt different behaviour without muscular co-contraction to fix posture. However, muscular tension associated with state anxiety [39, 97, 105] may have increased motor activity, which in turn increased both SA and SPL parameters. Finally, when all sensory inputs are available, state anxiety may modify the processing or integration of visual, vestibular and/or somatosensory inputs, whatever the trait anxiety level. In the present study, this led subjects to perform larger and faster body sways.

When vision is not available, results have shown a different effect of the anxiogenic condition on static balance performance according to the level of trait anxiety (see figure 4). Previous studies on this issue did not show any influence of state anxiety in the EC condition [71, 101], but none of them took the level of trait anxiety into account. In the present study, the VLTA group but not the ITA group increased the SA parameter and the SPL parameter in the anxiogenic condition. These results may provide initial insight into inter-individual trait anxiety differences in managing static balance without visual input. Similarly to the test performed with visual input, VLTA subjects have larger and faster body sways in the anxiogenic condition when vision is absent. In agreement with studies that proposed a relationship between sensory inputs and psychological states [5, 10, 42, 71, 101], we suggest that in this condition, state anxiety may influence vestibular input and/or somatosensory input. Moreover, the effects of state anxiety on static balance control do not seem to be dependent on whether visual input is present for VLTA subjects. In contrast, static balance control of ITA subjects may not be influenced by the anxiogenic condition when vision is absent.

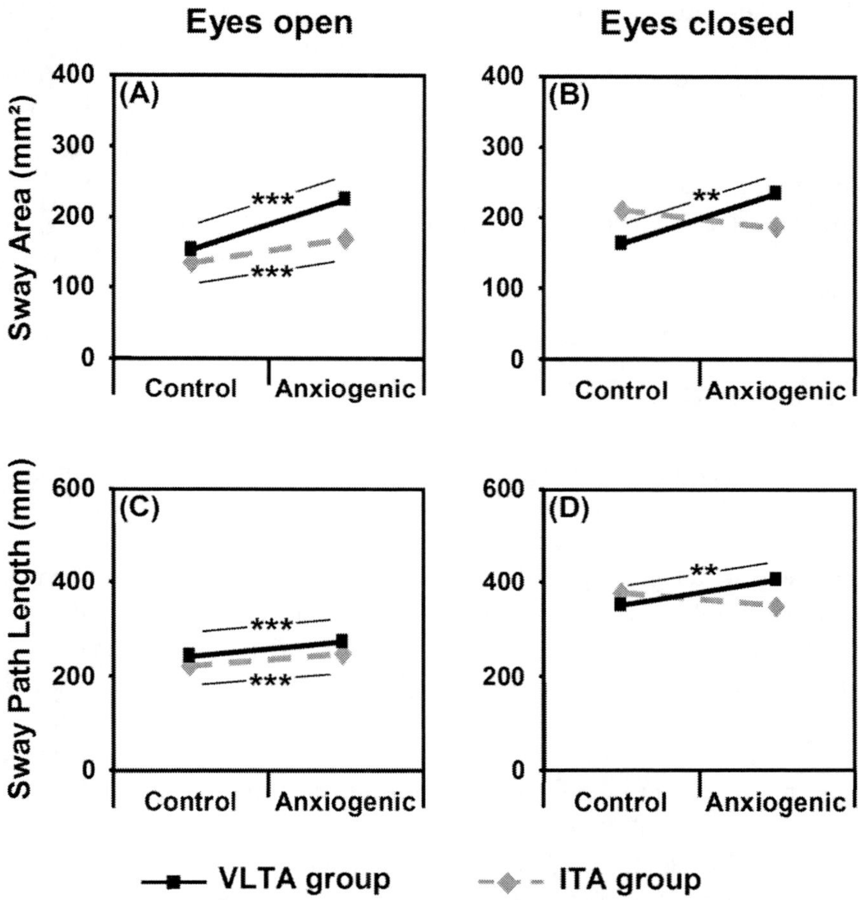

** $p < 0.01$, *** $p < 0.001$, significant difference between the control condition and the anxiogenic condition.

Figure 4. Differences in balance performances with eyes open (A, C) and eyes closed (B, D) for the sway area (mean in mm²; A, B) and the sway path length (mean in mm; C, D) between the control condition and the anxiogenic condition for the very low trait anxiety group and the intermediate trait anxiety group.

The lack of increase in SA and SPL parameters between control and anxiogenic conditions in the eyes-closed test in ITA subjects could suggest that moderate state-anxiety would not disturb the ability to use vestibular input and/or somatosensory input. Given the potential visual dependence of the ITA subjects that occur in the control condition, we hypothesized that the state anxiety induced -that remains moderate- could not be sufficient to cause additional perturbation. The main discrepancy between VLTA and ITA groups is that VLTA subjects are less visually dependent than ITA subjects, and so, more sensitive to the effects of moderate state anxiety in the EC test. These results are in concordance with those of previous studies that have demonstrated that phobic postural vertigo patients [41] or elderly adults with postural trait anxiety, characterized by high visual field dependence [23, 24] show poor balance performances in normal conditions compared to healthy subjects and young subjects respectively, but are less vulnerable to anxiogenic conditions.

Conclusion

Our results provide further evidence that state and trait anxiety could affect sensory inputs involved in balance performance. When vision is available, whatever the trait anxiety level, state anxiety could modify the processing or integration of visual, vestibular and/or somatosensory inputs. When vision is absent, the precision of static balance control in individuals with an intermediate level of trait anxiety, characterized by a higher visual dependence, is altered. Furthermore, moderate state anxiety would disturb the ability of individuals with a lower trait anxiety, but not that of individuals with intermediate trait anxiety, to use vestibular and/or somatosensory inputs. These results confirm the interdependence of trait anxiety, state anxiety and static balance control. It should be interesting that future research concerning multisensory integration in balance control incorporates psychological states including intrinsic predispositions in order to obtain a comprehensive assessment of balance control.

General Conclusion

Our work attempts to explore the influence of state and trait anxiety on multisensory integration involved in various sensorimotor tasks. The results of these studies have mainly evidenced that according to the trait anxiety level, state anxiety may have distinct influences on attentional processes underlying multisensory integration. We have suggested that the allocation of attentional resources toward sensory modalities could depend on the personal relevance of stimuli, which appears different between groups of low and intermediate trait anxiety level. Indeed, contrary to individuals with low level of trait anxiety, individuals with higher level of trait anxiety appear to have a greater visual dependence when performing sensorimotor tasks.

These results should be considered on the basis of previous studies in this research area. Most of the studies considering the relationships between emotions and crossmodal attention have used emotional stimuli (*e.g.*, [12, 13, 53, 58, 79]), and have highlighted some attentional biases that are specific to emotional processing. As far as anxious personality is concerned, previous studies have suggested that attentional biases could result from a disruption in the balance between the endogenous and the exogenous attention. Indeed, it has been shown that anxious subjects are automatically induced to direct their attention toward threatening stimuli [63], and have difficulties in voluntarily disengaging attention from them [32, 110]. The few studies that have used nonemotional stimuli in investigating the competition between endogenous and exogenous attention have also suggested that anxious individuals have an enhanced exogenous attentional system and that they are attracted to salient stimuli regardless of emotionality [64, 65].

It should be noted that in our studies, the experimental design is quite different. Indeed, subjects were asked to perform experimental tasks that require the processing of various modalities. There were neither task-irrelevant stimuli (*e.g.* distractor), nor salient stimuli (*e.g.* abrupt or high-contrast stimulations), nor emotional stimuli (*e.g.* threatening stimuli). Moreover, given our experimental design, we cannot assert that subjects have voluntarily or consciously directed their attention toward a specific modality. Our results rather suggest that the allocation of attentional resources was involuntarily and unconsciously conducted, depending on the subject's trait anxiety level. This mechanism, linked to the personality, could be considered as a filter on our sensory processing.

Such results seem to highlight that certain modalities could be specifically relevant according to the individual's personality. It should be noted that previous study has suggested a specific involvement of the amygdala, which is particularly associated with anxious temperament [83], in the appraisal of relevant stimuli [86]. Besides, it has been shown that the thalamus, highly linked to the amygdala, could direct attention to one sensory modality rather than another [61, 94, 95]. By directing attentional resources toward the personally relevant modality, these two structures, which are involved in the emotional system, might therefore play a role in the behavioural differences in sensory modality processing between individuals with contrasted trait anxiety. The assumption that the individual may have personal priorities in the attentional processing of sensory inputs is consistent with a recent study that has revealed a neural mechanism for the construction of a priority map involved in the selection of the important stimuli for attention [66].

The interest of this emerging research is to understand the role of personality in the influence of affective states on multisensory processing underlying action, in order to allow better adaptation processes of man that interact with environment. However, the dynamic relationship between personality traits, affective states and the weight of each sensory modality involved in sensorimotor tasks remains complex and ambiguous. Therefore, further data are needed to determine thoroughly the mechanisms by which the allocation of attentional resources may be inhibited, shifted or directed toward specific modalities.

References

[1] Abraini, JH; Bouquet, C; Joulia, F; Nicolas, M; Kriem, B. Cognitive performance during a simulated climb of mount everest: implications for brain function and central adaptive processes under chronic hypoxic stress. *Pflugers Arch*, 1998, 436, 553-9.

[2] Adkin, AL; Frank, JS; Carpenter, MG; Peysar, GW. Postural control is scaled to level of postural threat. *Gait Posture*, 2000, 12, 87-93.

[3] Alais, D; Newell, FN; Mamassian, P. Multisensory processing in review: from physiology to behaviour. *Seeing Perceiving*, 2010, 23, 3-38.

[4] Azais, F. [Cognitive experimental approach to anxiety disorders]. *Encephale*, 1995, 21, 597-607.

[5] Balaban, CD; Porter, JD. Neuroanatomic substrates for vestibulo-autonomic interactions. *J Vestib Res*, 1998, 8, 7-16.

[6] Balaban, CD; Thayer, JF. Neurological bases for balance-anxiety links. *J Anxiety Disord*, 2001, 15, 53-79.

[7] Bolmont, B; Thullier, F; Abraini, JH. Relationships between mood states; performances in reaction time, psychomotor ability, and mental efficiency during a 31-day gradual decompression in a hypobaric chamber from sea level to 8848 m equivalent altitude. *Physiol Behav*, 2000, 71, 469-76.

[8] Bolmont, B; Abraini, JH. State-anxiety and low moods: evidence for a single concept. *Physiol Behav*, 2001, 74, 421-4.

[9] Bolmont, B; Bouquet, C; Thullier, F. Relationships of personality traits with performance in reaction time, psychomotor ability, and mental efficiency during a 31-day simulated climb of Mount Everest in a hypobaric chamber'. *Percept Mot Skills*, 2001, 92, 1022-30.

[10] Bolmont, B; Gangloff, P; Vouriot, A; Perrin, PP. Mood states and anxiety influence abilities to maintain balance control in healthy human subjects. *Neurosci Lett*, 2002, 329, 96-100.

[11] Bolmont, B. Role and influence of moods including anxiety on motor control. In Clark AV (Ed.). *Causes, Role and influence of Moods States*, Nova Science Publisher, New York, 2005, 57-73.

[12] Brosch, T; Grandjean, D; Sander, D; Scherer, KR. Behold the voice of wrath: cross-modal modulation of visual attention by anger prosody. *Cognition*, 2008, 106, 1497-503.

[13] Brosch, T; Grandjean, D; Sander, D; Scherer, KR. Cross-modal emotional attention: emotional voices modulate early stages of visual processing. *J Cogn Neurosci*, 2009, 21, 1670-9.

[14] Brown, LA; Polych, MA; Doan, JB. The effect of anxiety on the regulation of upright standing among younger and older adults. *Gait Posture*, 2006, 24, 397-405.

[15] Busse, L; Roberts, KC; Crist, RE; Weissman, DH; Woldorff, MG. The spread of attention across modalities and space in a multisensory object. *Proc Natl Acad Sci U S A*, 2005, 102, 18751-6.

[16] Carpenter, MG; Frank, JS; Silcher, CP. Surface height effects on postural control: a hypothesis for a stiffness strategy for stance. *J Vestib Res*, 1999, 9, 277-86.

[17] Carpenter, MG; Frank, JS; Silcher, CP; Peysar, GW. The influence of postural threat on the control of upright stance. *Exp Brain Res*, 2001, 138, 210-8.

[18] Colavita, FB. Human sensory dominance. Percept. *Psychophys*., 1974, 16, 409-412.

[19] Deutsch, JA; Deutsch, D. Attention: some theoretical considerations. *Psychol Rev*, 1963, 70, 80-90.

[20] Doumas, M; Krampe, RT. Adaptation and reintegration of proprioceptive information in young and older adults' postural control. *J Neurophysiol*, 2010, 104, 1969-77.

[21] Driver, J; Spence, C. Crossmodal attention. *Curr Opin Neurobiol*, 1998, 8, 245-53.

[22] Driver, J; Frackowiak, RS. Neurobiological measures of human selective attention. *Neuropsychologia*, 2001, 39, 1257-62.

[23] Eikema, DJ; Hatzitaki, V; Tzovaras, D; Papaxanthis, C. Age-dependent modulation of sensory reweighting for controlling posture in a dynamic virtual environment. *Age* (Dordr), 2012, 34, 1381-92.

[24] Eikema, DJ; Hatzitaki, V; Konstantakos, V; Papaxanthis, C. Elderly adults delay proprioceptive reweighting during the anticipation of collision avoidance when standing. *Neuroscience*, 2013, 234, 22-30.

[25] Eklund, G. General features of vibration-induced effects on balance. *Ups J Med Sci*, 1972, 77, 112-24.

[26] Era, P; Heikkinen, E. Postural sway during standing and unexpected disturbance of balance in random samples of men of different ages. *J Gerontol*, 1985, 40, 287-95.

[27] Etnyre, B; Kinugasa, T. Postcontraction influences on reaction time. *Res Q Exerc Sport*, 2002, 73, 271-81.

[28] Eysenck, MW. *Anxiety: The cognitive perspective*. Lawrence Erlbaum; 1992.

[29] Eysenck, MW; Derakshan, N; Santos, R; Calvo, MG. Anxiety; cognitive performance: attentional control theory. *Emotion*, 2007, 7, 336-53.

[30] Fitzpatrick, R; McCloskey, DI. Proprioceptive, visual; vestibular thresholds for the perception of sway during standing in humans. *J Physiol*, 1994, 478.1, 173-86.

[31] Fowler, B; Lindeis, AE. The effects of hypoxia on auditory reaction time; P300 latency. *Aviat Space Environ Med*, 1992, 63, 976-81.

[32] Fox, E; Russo, R; Bowles, R; Dutton, K. Do threatening stimuli draw or hold visual attention in subclinical anxiety? *J Exp Psychol Gen*, 2001, 130, 681-700.

[33] Furman, JM; Jacob, RG. A clinical taxonomy of dizziness; anxiety in the otoneurological setting. *J Anxiety Disord*, 2001, 15, 9-26.

[34] Hainaut, JP; Bolmont, B. Effects of mood states and anxiety as induced by the video-recorded stroop color-word interference test in simple response time tasks on reaction time and movement time. *Percept Mot Skills*, 2005, 101, 721-9.

[35] Hainaut, JP; Bolmont, B. Moderate state-anxiety differently modulates visual and auditory response times in normal- and very low trait-anxiety subjects. *Neurosci Lett*, 2006, 395, 129-32.

[36] Hainaut, JP; Monfort, V; Bolmont, B. Trait-anxiety dependence of movement time performance in a bimodal choice task in subjects exposed to moderate anxiogenic conditions. *Neurosci Lett*, 2006, 409, 196-9.

[37] Hainaut, JP; Caillet, G; Lestienne, FG; Bolmont, B. The role of trait anxiety on static balance performance in control and anxiogenic situations. *Gait Posture*, 2011, 33, 604-8.

[38] Hazlett, RL; McLeod, DR; Hoehn-Saric, R. Muscle tension in generalized anxiety disorder: elevated muscle tonus or agitated movement? *Psychophysiology*, 1994, 31, 189-95.

[39] Hoehn-Saric, R; Hazlett, RL; Pourmotabbed, T; McLeod, DR. Does muscle tension reflect arousal? Relationship between electromyographic and electroencephalographic recordings. *Psychiatry Res*, 1997, 71, 49-55.

[40] Hoehn-Saric, R; McLeod, DR. Anxiety and arousal: physiological changes; their perception. *J Affect Disord*, 2000, 61, 217-24.

[41] Holmberg, J; Tjernstrom, F; Karlberg, M; Fransson, PA; Magnusson, M. Reduced postural differences between phobic postural vertigo patients and healthy subjects during a postural threat. *J Neurol*, 2009, 256, 1258-62.

[42] Jacob, RG; Furman, JMR; Clark, DB; Durrant, JD; Balaban, CD. *Psychogenic dizziness*. In Sharp JA and Barber HO (Eds.). The Vestibulo-ocular reflex and vertigo, Raven Press, New York, 1993, 305-315.

[43] Jacob, RG; Furman, JM; Durrant, JD; Turner, SM. Panic, agoraphobia,; vestibular dysfunction. *Am J Psychiatry*, 1996, 153, 503-12.

[44] Jeka, J; Allison, L; Saffer, M; Zhang, Y; Carver, S; Kiemel, T. Sensory reweighting with translational visual stimuli in young and elderly adults: the role of state-dependent noise. *Exp Brain Res*, 2006, 174, 517-27.

[45] Jeka, JJ; Oie, KS; Kiemel, T. Asymmetric adaptation with functional advantage in human sensorimotor control. *Exp Brain Res*, 2008, 191, 453-63.

[46] Jensen, AR; Munro, E. Reaction time, movement time, and intelligence. *Intelligence*, 1979, 3, 121-126.

[47] Jones, JG; Hardy, L. The effects of anxiety upon psychomotor performance. *J Sports Sci*, 1988, 6, 59-67.

[48] Kapoula, Z; Le, TT. Effects of distance and gaze position on postural stability in young; old subjects. *Exp Brain Res*, 2006, 173, 438-45.

[49] Kimura, K; Imanaka, K; Kita, I. The effects of different instructions for preparatory muscle tension on simple reaction time. *Hum Mov Sci*, 2002, 21, 947-60.

[50] Klein, RM. Attention and visual dominance: a chronometric analysis. *J Exp Psychol Hum Percept Perform*, 1977, 3, 365-78.

[51] Koelewijn, T; Bronkhorst, A; Theeuwes, J. Attention and the multiple stages of multisensory integration: A review of audiovisual studies. *Acta Psychol (Amst)*, 2010, 134, 372-84.
[52] Kogan, E; Lidor, R; Bart, O; Bar-Haim, Y; Mintz, M. Comorbidity between balance and anxiety disorders: verification in a normal population. *J Psychol*, 2008, 142, 601-13.
[53] Koizumi, A; Tanaka, A; Imai, H; Hiramatsu, S; Hiramoto, E; Sato, T; de Gelder, B. The effects of anxiety on the interpretation of emotion in the face-voice pairs. *Exp Brain Res*, 2011, 213, 275-82.
[54] Le, TT; Kapoula, Z. Role of ocular convergence in the Romberg quotient. *Gait Posture*, 2008, 27, 493-500.
[55] Leite, JR; Seabra Mde, L; Sartori, VA; Andreatini, R. The video-recorded Stroop Color-Word Test as a new model of experimentally-induced anxiety. *Prog Neuropsychopharmacol Biol Psychiatry*, 1999, 23, 809-22.
[56] Lewkowicz, DJ; Ghazanfar, AA. The emergence of multisensory systems through perceptual narrowing. *Trends Cogn Sci*, 2009, 13, 470-8.
[57] Linton, SJ; Gotestam, KG. Relations between pain, anxiety, mood and muscle tension in chronic pain patients. A correlation study. Psychother Psychosom, 1985, 43, 90-5.
[58] Maiworm, M; Bellantoni, M; Spence, C; Roder, B. When emotional valence modulates audiovisual integration. *Atten Percept Psychophys*, 2012, 74, 1302-11.
[59] Martens, R. Arousal and motor performance. *Exercise and Sport Sciences Reviews*, 1974, 2, 155-188.
[60] Mathews, A. Why worry? The cognitive function of anxiety. *Behav Res Ther*, 1990, 28, 455-68.
[61] McCormick, DA; Bal, T. Sensory gating mechanisms of the thalamus. *Curr Opin Neurobiol*, 1994, 4, 550-6.
[62] Miyake, A; Friedman, NP; Emerson, MJ; Witzki, AH; Howerter, A; Wager, TD. The unity and diversity of executive functions and their contributions to complex "Frontal Lobe" tasks: a latent variable analysis. *Cogn Psychol*, 2000, 41, 49-100.
[63] Mogg, K; Bradley, BP. Some methodological issues in assessing attentional biases for threatening faces in anxiety: a replication study using a modified version of the probe detection task. *Behav Res Ther*, 1999, 37, 595-604.

[64] Moriya, J; Tanno, Y. Competition between endogenous and exogenous attention to nonemotional stimuli in social anxiety. *Emotion*, 2009, 9, 739-43.

[65] Moriya, J; Sugiura, Y. Impaired attentional disengagement from stimuli matching the contents of working memory in social anxiety. *PLoS One*, 2012, 7, e47221.

[66] Mysore, SP; Knudsen, EI. A shared inhibitory circuit for both exogenous; endogenous control of stimulus selection. *Nat Neurosci*, 2013, 16, 473-8.

[67] Nashner, LM; Shupert, CL; Horak, FB. Head-trunk movement coordination in the standing posture. *Prog Brain Res*, 1988, 76, 243-51.

[68] Noteboom, JT; Barnholt, KR; Enoka, RM. Activation of the arousal response and impairment of performance increase with anxiety and stressor intensity. *J Appl Physiol*, 2001, 91, 2093-101.

[69] Noteboom, JT; Fleshner, M; Enoka, RM. Activation of the arousal response can impair performance on a simple motor task. *J Appl Physiol*, 2001, 91, 821-31.

[70] O'Connor, KW; Loughlin, PJ; Redfern, MS; Sparto, PJ. Postural adaptations to repeated optic flow stimulation in older adults. *Gait Posture*, 2008, 28, 385-91.

[71] Ohno, H; Wada, M; Saitoh, J; Sunaga, N; Nagai, M. The effect of anxiety on postural control in humans depends on visual information processing. *Neurosci Lett*, 2004, 364, 37-9.

[72] Ouaknine, M, *Posture and vision, 4th symposium of CEOPS*, Vol. 4, Toulouse France, 2006, 18-22.

[73] Oxendine, JB. *Emotional arousal and motor performance*, Quest n°13, 1970, 23-32.

[74] Palmer, TD; Ramsey, AK. *The function of consciousness in multisensory integration*. Cognition, 2012, 125, 353-64.

[75] Paulus, W; Straube, A; Brandt, TH. Visual postural performance after loss of somatosensory and vestibular function. *J Neurol Neurosurg Psychiatry*, 1987, 50, 1542-5.

[76] Peretti, CS. [Anxiety; cognition disorders]. *Encephale*, 1998, 24, 256-9.

[77] Peterka, RJ. Sensorimotor integration in human postural control. *J Neurophysiol*, 2002, 88, 1097-118.

[78] Peterka, RJ; Loughlin, PJ. Dynamic regulation of sensorimotor integration in human postural control. *J Neurophysiol*, 2004, 91, 410-23.

[79] Poliakoff, E; Miles, E; Li, X; Blanchette, I. The effect of visual threat on spatial attention to touch. *Cognition*, 2007, 102, 405-14.

[80] Posner, MI; Nissen, MJ; Klein, RM. Visual dominance: an information-processing account of its origins and significance. *Psychol Rev*, 1976, 83, 157-71.

[81] Posner, MI. Orienting of attention. *Q J Exp Psychol*, 1980, 32, 3-25.

[82] Raglin, JS. Anxiety and sport performance. *Exerc Sport Sci Rev*, 1992, 20, 243-74.

[83] Rauch, SL; Shin, LM; Wright, CI. Neuroimaging studies of amygdala function in anxiety disorders. *Ann N Y Acad Sci*, 2003, 985, 389-410.

[84] Redfern, MS; Jennings, JR; Martin, C; Furman, JM. Attention influences sensory integration for postural control in older adults. *Gait Posture*, 2001, 14, 211-6.

[85] Redfern, MS; Jennings, JR; Mendelson, D; Nebes, RD. Perceptual inhibition is associated with sensory integration in standing postural control among older adults. *J Gerontol B Psychol Sci Soc Sci*, 2009, 64, 569-76.

[86] Sander, D; Grafman, J; Zalla, T. The human amygdala: an evolved system for relevance detection. *Rev Neurosci*, 2003, 14, 303-16.

[87] Shapiro, KL; Lim, A. The impact of anxiety on visual attention to central and peripheral events. *Behav Res Ther*, 1989, 27, 345-51.

[88] Silva, FT; Leite, JR. Physiological modifications; increase in state anxiety in volunteers submitted to the Stroop Color-Word Interference Test: A preliminary study. *Physiol Behav*, 2000, 70, 113-8.

[89] Sklare, DA; Konrad, HR; Maser, JD; Jacob, RG. Special issue on the interface of balance disorders; anxiety: an introduction and overview. *J Anxiety Disord*, 2001, 15, 1-7.

[90] Spence, C. Multisensory attention and tactile information-processing. *Behav Brain Res*, 2002, 135, 57-64.

[91] Spielberger, CD. *Anxiety: Current trends in theory and research.* Academic Press; 1972.

[92] Spielberger, CD; Gorsuch, RL; Lushene, R; Vagg, PR; Jacobs, GA. *Manual for the State-Trait Anxiety Inventory (Form Y) (Self-Evaluation Questionnaire)*. Palo Alto, CA: Consulting Psychologists Press, 1983.

[93] Spijkers, WA. The relation between response-specificity, S-R compatibility, foreperiod duration and muscle-tension in a target aiming task. *Acta Psychol* (Amst), 1990, 75, 261-77.

[94] Sturm, W; de Simone, A; Krause, BJ; Specht, K; Hesselmann, V; Radermacher, I; Herzog, H; Tellmann, L; Muller-Gartner, HW; Willmes, K. Functional anatomy of intrinsic alertness: evidence for a

fronto-parietal-thalamic-brainstem network in the right hemisphere. *Neuropsychologia*, 1999, 37, 797-805.

[95] Sturm, W; Longoni, F; Fimm, B; Dietrich, T; Weis, S; Kemna, S; Herzog, H; Willmes, K. Network for auditory intrinsic alertness: a PET study. *Neuropsychologia*, 2004, 42, 563-8.

[96] Talsma, D; Senkowski, D; Soto-Faraco, S; Woldorff, MG. The multifaceted interplay between attention and multisensory integration. *Trends Cogn Sci*, 2010, 14, 400-10.

[97] Teixeira-Silva, F; Prado, GB; Ribeiro, LC; Leite, JR. The anxiogenic video-recorded Stroop Color-Word Test: psychological and physiological alterations; effects of diazepam. *Physiol Behav*, 2004, 82, 215-30.

[98] Tulen, JH; Moleman, P; van Steenis, HG; Boomsma, F. Characterization of stress reactions to the Stroop Color Word Test. *Pharmacol Biochem Behav*, 1989, 32, 9-15.

[99] Viaud-Delmon, I; Ivanenko, YP; Berthoz, A; Jouvent, R. Adaptation as a sensorial profile in trait anxiety: a study with virtual reality. *J Anxiety Disord*, 2000, 14, 583-601.

[100] Visser, JE; Carpenter, MG; van der Kooij, H; Bloem, BR. The clinical utility of posturography. *Clin Neurophysiol*, 2008, 119, 2424-36.

[101] Wada, M; Sunaga, N; Nagai, M. Anxiety affects the postural sway of the antero-posterior axis in college students. *Neurosci Lett*, 2001, 302, 157-9.

[102] Waersted, M; Bjorklund, RA; Westgaard, RH. Shoulder muscle tension induced by two VDU-based tasks of different complexity. *Ergonomics*, 1991, 34, 137-50.

[103] Waersted, M; Bjorklund, RA; Westgaard, RH. The effect of motivation on shoulder-muscle tension in attention-demanding tasks. *Ergonomics*, 1994, 37, 363-76.

[104] Willmann, M; Bolmont, B. The trapezius muscle uniquely lacks adaptive process in response to a repeated moderate cognitive stressor. *Neurosci Lett*, 2012, 506, 166-9.

[105] Willmann, M; Langlet, C; Hainaut, JP; Bolmont, B. The time course of autonomic parameters and muscle tension during recovery following a moderate cognitive stressor: dependency on trait anxiety level. *Int J Psychophysiol*, 2012, 84, 51-8.

[106] Woollacott, MH; Shumway-Cook, A; Nashner, LM. Aging; posture control: changes in sensory organization; muscular coordination. *Int J Aging Hum Dev*, 1986, 23, 97-114.

[107] Yardley, L; Britton, J; Lear, S; Bird, J; Luxon, LM. Relationship between balance system function; agoraphobic avoidance. *Behav Res Ther*, 1995, 33, 435-9.
[108] Yardley, L; Owen, N; Nazareth, I; Luxon, L. Prevalence and presentation of dizziness in a general practice community sample of working age people. *Br J Gen Pract*, 1998, 48, 1131-5.
[109] Yerkes, RM; Dodson, JD. The relation of strength of stimulus to rapidity of habit information. *J Compar Neurol Psych*, 1908, 18, 459-482.
[110] Yiend, J; Mathews, A. Anxiety and attention to threatening pictures. *Q J Exp Psychol A*, 2001, 54, 665-81.

In: New Developments in Sensory Processing … ISBN: 978-1-62808-395-8
Editors: M. Voisin and R. Brunel

Chapter IV

Regulatory Disorders of Sensory Processing: The Role of Sensory Modulation Deficits in a Diagnostic in Early Childhood

Ruth Pérez-Robles[1,*], Eduardo Doval[2], M. Claustre Jané[1] and Pedro Caldeira da Silva[3]

[1]Department of Clinical and Health Psychology.
Universitat Autònoma de Barcelona, Barcelona, Spain
[2]Departament of Psychobiology and Methodology of Health Sciences.
Universitat Autònoma de Barcelona, Barcelona, Spain
[3]Infancy And Early Childhood Unit, Department of Child and Adolescent Psychiatry. CHLC-Hospital Dona Estefânia, Lisbon, Portugal

Abstract

Regulation Disorders of Sensory Processing (RDSP) is a clinical diagnosis in the *Diagnostic Classification of Mental Health and Development Disorders of Infancy and Early Childhood Revised edition:*

* Corresponding author: Ruth Pérez Robles. Departament of Clinical and Health Psychology, Universitat Autònoma de Barcelona, 08193 Bellaterra, Barcelona, Spain. E-mail: Ruth.Perez@uab.cat/ ruthprobles@gmail.com.

DC:0-3R (Zero to Three, 2005). Data supporting RDSP as a distinct and valid disorder are scant. For this reason it was not included in the DSM diagnostic classification system. The aim of this chapter is to contribute to the validation of the sensory criteria of RDSP summarizing the findings from previous studies from the authors. In those studies it was analyzed in clinical samples: (1) the severity of sensory modulation deficits in children with RDSP; (2) the associations between sensory and behavioral symptoms; and (3) the specific role of sensory modulation deficits on the diagnosis of RDSP. Method: The samples were composed by toddlers (18-36 months-old) who were diagnosed by the clinician based on DC:0-3R diagnostic criteria. Toddlers were classified into two groups: those with RDSP and those with ''other diagnoses in Axis I/II of the DC:0-3R''(OD3R). Independently of the clinical process, parents completed the Infant Toddler Sensory Profile (as a checklist for sensory symptoms) and the Achenbach Behavior Checklist for ages 1 ½ –5 (CBCL 1 ½-5). The scores from the two groups were compared. Results: The results revealed that the RDSP group had more severe sensory modulation deficits (and more severe specific behavioral symptoms); stronger, although not significant, associations between most sensory and behavioral symptoms; and a significant sensory modulation deficits effect. Conclusion: The results confirmed the hypotheses, demonstrating the discriminative capacities of RDSP criteria versus OD3R, the independence of both sensory and behavioral criteria and the significant effect of sensory modulation deficits on RDSP diagnostic. The results also revealed convergent validity between the instruments and the diagnostic criteria. These findings suggest the importance of identifying RDSP in order to enhance the sensory capacities of children and to prevent clinical conditions at later stages.

Introduction

In the field of occupational therapy during the sixties in the USA, the occupational therapist Jean Ayres documented for the first time the interindividual differences of sensory processing in young children. The observation of such differences resulted in the concept of *"sensory integration dysfunction"* (SID) characterized by a decreased ability of the brain to organize sensory information as it comes in through the senses [1]. To treat these dysfunctions, the scientific discipline of "Sensory Integration" was created [2].

The sixties also witnessed the emergence of the first theoretical models regarding child temperament. Some of these models had already considered "sensory reactivity" and "self-regulation ability" to be crucial elements of temperament, highlighting the importance of evaluating these characteristics in early childhood [3]. However, these concepts were not approached from the perspective of psychopathology until the nineties. At that time, research began to focus on elucidating the connection between relationships in sensory processing, the development of emotional and social maturity, and the emergence of behavioral and emotional disorders in early childhood.

In this context, several studies were conducted for the first time with infants aged over 6 months who were already displaying "difficulties in regulating physiological, sensory, attentional, motor, or affective processes and maintaining a calm, alert, or affectively positive state" [4, 5, 6]. Some of these studies had demonstrated that a large proportion of these infants presented patterns of reactivity and habituation to sensory stimuli that were distinct from the usual responses [5]. Some of them have also indicated that 95% of infants who exhibit moderate regulatory problems within their first year of life are later diagnosed at the age of 3 with delays in motor, language, and cognitive development or relational disorders [4, 6].

These types of results led to the creation of the diagnosis of *Regulatory Disorders* (RD) in the field of child psychology, which appeared for the first time in the *Diagnostic Classification for Infant and Early Childhood Mental Health and Developmental Disorders*, DC:0-3 [7]. This diagnostic classification was meant to complement the *Diagnostic and Statistical Manual of Mental Disorders*, DSM-IV [8], and the *International Classification of Diseases*, ICD-10 [9], for the first 3 years of life. In the revised edition of DC:0-3R [10], the term *Regulatory Disorders* was revised and changed to *Regulatory Disorders of Sensory Processing* (RDSP) to highlight the significance of the alteration of sensory processing in this diagnosis.

The only European epidemiological study, which was conducted in Denmark, indicated that 7% of the children aged over 18 months in the general population met the criteria described in this diagnosis [11].

The Regulatory Disorders of Sensory Processing diagnosis

RDSP are currently characterized by ''difficulties in regulating emotions and behaviors, as well as motor abilities, in response to sensory stimulation that lead to impairment in development and functioning. These behaviors occur in multiple relationships and contexts'' [10]. These difficulties must be intense enough to interfere with daily life and emerge in different contexts and different relations. This disorder can be diagnosed in infants as young as 6 months of age.

In the first edition of DC:0-3 [7], there were 2 criteria for an RD diagnosis: 1) alterations in sensory processing or difficulties in modulating the motor response; and 2) a characteristic behavioral pattern of the child [12]. In the current DC:0-3R [10], a third criterion was added to diagnose RDSP: the presence of motor symptoms was distinguished from the other two criteria.

The diagnosis of RDSP is divided into 3 types and 2 subtypes [10]: Hypersensitive (fearful/cautious or negative/defiant), Hyposensitive/under-responsive, Sensory stimulation-seeking/impulsive.

Types

The types are based on the sensory threshold of the child (high or low) and on the self-regulation strategies (active or passive) that the child uses to achieve homeostasis. The symptoms can vary depending on age. The sensory processing alterations can be noted in the following functions: visual, auditory, tactile, vestibular, proprioceptive, oral sensitivity (taste or texture), smell, motor planning, muscle tone, and attention. All 3 types of RDSP in the DC:0-3R are characterized by the extreme responses of the child, which interfere with self-care, play, and learning activities. The 3 types are described as follows [10, pp.36-38]:

Type I. ''Hypersensitive'' describes children with a low sensory threshold. Depending on the self-regulatory strategies of these children, those diagnosed with Type I are classified into the following subtypes:

- Subtype A—''Fearful/cautious''. The sensory reactivity patterns of these children are characterized by an over-reactivity to sensory stimuli; they appear frightened by nearby sounds, movements, or visual stimuli. The motor patterns are characterized by limited sensory-motor play and less exploration than expected according to age. The behavioral patterns are characterized by excessive cautiousness, inhibition, and fearfulness.
- Subtype B—''Negative/defiant''. The sensory reactivity patterns of these children are identical to those of subtype A. The motor patterns of these children are characterized by avoidance

behaviors or being slow to engage in new experiences. The behavioral patterns include the following: negativistic behavior; controlling behaviors; defiance; and in general, aggression only when provoked.

Type II. ''Hyposensitive/Under-responsive'' children have a high sensory threshold. The sensory reactivity patterns of the children are characterized by under-reactivity to sounds, movement, smell, taste, touch, and/or proprioception. The motor patterns include the following: limited exploration; a restricted play repertoire; lethargy; poor motor planning; and/or searching for a specific sensory input. The behavioral patterns are characterized by an apparent lack of interest in exploring the properties of objects, playing challenging games, or engaging in social interactions, an apathetic appearance, fatigability, and/or inattentiveness.

Type III. ''Sensory Stimulation-Seeking/Impulsive'' children also have a high sensory threshold, but these children use active self-regulation strategies. The sensory reactivity patterns are characterized by the child's craving for high-intensity sensory stimuli. The motor patterns are characterized by a strong need for motor discharge, diffuse impulsivity, and/or accident proneness without clumsiness. The behavioral patterns can include seeking constant contact with people and objects, recklessness, and disorganized behavior as a consequence of sensory stimulation.

These types are similar to those suggested by Dunn (2002) in the field of occupational therapy and are associated with a similar diagnosis called *Sensory Modulation Dysfunction* (SMD), which is characterized by an increased or decreased sensitivity to sensory stimuli [13].

Diagnostic Validity and Child Psychodiagnostic Classification Systems

It should be noted that the creation of child psychodiagnostic classification systems stemmed from two different approaches [14]. On the one hand, some classifications, such as the *Research Diagnostic Criteria-Preschool Age*, RDC-PA [15], used a *top-down* approach; i.e., starting from the diagnostic classifications for adults, the diagnostic criteria were modified for application in children. On the other hand, for the first version of the DC:0-3 [7] typology, a *bottom-up* approach was used; i.e., the specific diagnoses for early childhood were created based on the consensus of clinical perspectives and scientific knowledge regarding the development of infants and children of that age.

The current DC:0-3R [10] represents a combination of the two aforementioned approaches. The structure of the DC:0-3R is based on 5 diagnostic axes: Axis I describes the Clinical Disorder distributed into 7 main diagnostic categories; Axis II addresses the parent-child relational diagnoses; Axis III describes the medical conditions related to the retardation of development; and Axes IV and V provide complementary information related to the factors of psychosocial and family risk and regarding the socio-emotional development of the child. The DC:0-3R is now the most commonly diagnostic classification used in infancy and early childhood.

The approach to diagnosing RDSP is supported by only a few validation studies. This situation is rather common among the diagnoses specific to early childhood that were derived from clinical consensus, in contrast to the diagnoses based on the adaptation of adult criteria to children, such as Depression [16], Post-Traumatic Stress [17,18], Sleep Disorders [19, 20, 21, 22, 23], and Childhood Anorexia [24,25]. Authors such as Rescorla [26] have indicated that the diagnosis of RDSP has not followed the standard procedures for evaluating the reliability and validity of the diagnoses established in the sector of child psychopathology. These procedures should include samples of "cases" with various issues, as well as "controls" and validate the diagnosis with data obtained from independent instruments, which had not been used in the diagnostic process.

To be considered valid, a diagnosis must pass through a process leading to the conclusion that the characteristics it defines are authentic representations of evidence found in clinical practice and not mere clinical inferences [27, 28].

The validation model for psychiatric categories suggested by Robins and Guze [29] has had a great impact in the field of child psychopathology. This model consists of five steps, which were incorporated into the Feighner Criteria and into the Research Diagnostic Criteria, which have been the basis of the DSM and CIE systems since then. This model was adopted by Cantwell [31] for the validation of psychopathologic diagnoses and consists of the following phases for information collection: 1) clinical phenomenology; 2) psychosocial factors; 3) demographic factors; 4) biological factors; 5) genetic factors; 6) family environment factors; 7) natural evolution; and 8) the response to therapy [31].

The first phase, clinical phenomenology, includes the symptomatic description of the clinical framework, the delineation of its boundaries, and the identification of comorbidity with other diagnoses. This phase uses dimensional and category instruments. Once these aspects have been defined, many studies must be conducted to develop the following seven phases of the model. The fact that empirical pieces of evidence were still lacking regarding this first phase in relation to RDSP has led to doubts concerning the nature of this diagnosis, which in the end was not included in the DSM-V [32].

When validating the framework and the related diagnostic criteria, it is very important to use standardized instruments that are able to reveal evidence regarding both the converging validity related to the clinical opinion of the symptoms of the diagnosis and the discriminating validity related to other similar diagnoses [33].

Based on the previous considerations, the questions that were most frequently raised in the scientific literature regarding the validity of RDSP will be presented. Some of the questions are the focus of our study and will be presented more in detail.

Important Aspects Regarding the Validity of RDSP in the Scientific Literature

First Aspect: The Discriminative Ability of the Sensory Criterion of the RDSP Diagnosis

One of the most frequent aspects mentioned in the scientific literature has to do with the discriminative ability of the sensory and behavioral criteria of the RDSP diagnosis compared to other diagnoses in child psychiatry/psychology [34, 35, 36, 37]. Several studies have indicated phenotypic similarities of RDSP with other disorders, such as anxiety [38] or other affective problems [39], oppositional/ defiant problems [40], or autistic disorder spectrum [34], which led to questioning the differential diagnostic ability of the symptoms of RDSP compared with other diagnoses in child psychiatry/psychology.

The DC:0-3 framework highlights that if both behavioral and sensory symptoms were not present in the child, ‘‘the clinician should consider alternative classifications to distinguish this diagnosis from others that might have phenotypic similarities in Axis I or II’’ [7, 8].

To study the discriminative ability of the behavioral and sensory criteria of the RDSP diagnosis compared to other diagnoses of the DC:0-3R, the authors conducted a study [41] aimed at analyzing the presence and severity of such symptoms in 2 clinical samples of 60 infants aged 18 to 36 months. The members of the first group had a clinical diagnosis of RDSP, and the second group had other diagnoses of Axes I/II of the DC:0-3R (OD3R), such as Post-traumatic Stress Disorders, Affective Disorders, Adaptation Disorders, functional Sleep or Nutrition Disorders, or relational diagnoses.

Presence of Sensory Symptoms in Both Clinical Samples

In the first study [41], the hypothesis was that the presence of alterations in sensory processing would be more significant in the sample of children with RDSP than the children with OD3R. To validate this hypothesis, the sensory symptoms were assessed independently of the clinical process with the *Infant Toddler Sensory Profile,* ITSP [42, 43]. This questionnaire includes 48 items that assess behaviors related to the sensory processing of children who are between 0 and 36 months of age. These items are assessed using a 5-point scale from 1 (almost always) to 5 (hardly ever). This instrument produces results according to sensory area and sensory pattern. These results include ‘‘low registration’’, ‘‘sensation seeking’’, ‘‘sensory sensitivity’’, and ‘‘sensation avoidance’’. The sum of the ‘‘sensory sensitivity’’ and ‘‘sensation avoidance’’ pattern scores compose the ‘‘low sensory threshold’’ category. The cut-off scores were determined in the following manner: ‘‘Typical performance’’ corresponded to scores at or between ±1 standard deviation (SD) from the mean of children without disabilities, ‘‘probable difference’’ corresponded to scores between ±1 SD and ±2 SD away from the mean, and ‘‘definite difference’’ corresponded to scores ±2 SD away from the mean. In this study [41], scores greater than 1 SD away from the mean denoted a deficit in sensory modulation (DC:0-3R criteria).

The results of the study validated the hypothesis because all of the children who were clinically diagnosed with RDSP presented alterations in at least one sensory field according to the ITSP [42], with scores varying at least 1 SD of the mean (above or below). This variation did not occur in the OD3R group, where 21% of the children did not show any alteration. However, the differences were only significant among the groups with the presence of two or more sensory areas affected.

Severity of the Sensory Symptoms

To analyze the severity of the sensory symptoms, another study with a sample of 74 children of the same age range was conducted in addition to the one discussed in the previous paragraph [44]. The hypothesis was that the sensory symptoms would be more serious in the RDSP clinical group than in OD3R clinical group. In the first study [41], the differences among the groups were significant for the hearing function. In the second study, they were significant in the visual and oral functions as well as the hearing function. These differences in results can be explained either by the size of the samples or by the nature of the variable "severity of the symptoms". In the first study, this variable was considered in a dichotomous way (having sensory deficits or not) in line with the criterion of the ITSP manual [42], while in the second study, this variable was considered in a quantitative way based on the score for each symptom.

In the two previous works [41, 44], significant differences were found in 2 out of 4 sensory patterns (one pattern characterized by a high sensory threshold, and the other pattern characterized by a low sensory threshold). Differences were also found in the low sensory threshold category. These results prove that both patterns, sensory under-reactivity and sensory over-reactivity, could entail risk factors in the child's development, as mentioned in various studies [46-56]. Despite the small sample size, these results nevertheless illustrate characteristics of various sensory patterns within the RDSP framework.

Second Aspect: The Relationship between Sensory and Behavioral Symptoms

Another issue regarding the validity of the RDSP diagnosis commonly found in the scientific literature is whether the sensory and behavioral symptoms characteristic of RDSP actually refer to the same concept but simply from different perspectives, for instance, from a child psychiatry/psychology approach or an occupational therapy approach [39, 48, 57, 58]. Some studies that examined the co-occurrence of sensory regulation dysfunction and preschool psychiatric disorders have concluded that 33–63% of the children meeting the criteria for sensory deregulation also had a psychiatric disorder [57]. This means that the relationship between the sensory and behavioral symptoms required analysis.

To address this issue, the second study from the authors [44] included a comparison of the association of the sensory and behavioral symptoms between the two clinical groups. The behavioral symptoms were assessed with the *Child Behavior Check-list* [59].

The hypotheses were as follows: 1) the sensory and behavioral symptoms would be independent entities; and 2) the association between the sensory and behavioral symptoms would be more significant in RDSP.

The associations between the sensory and behavioral symptoms were low to moderate, which seems to favor of the first hypothesis. This result reflects a certain independence between the two types of symptoms (sensory and behavioral), which argues against the idea that the symptoms reflect the same concept approached from two different theoretical frameworks. Similar results have been found in other studies [45, 57, 58].

The results did not validate the second hypothesis because no significant differences appeared between the two groups analyzed (RDSP and OD3R) regarding the relationship between the behavioral and sensory symptoms. Only a few associations tended to confirm the hypothesis; for example, there was an association between the behavioral avoidance symptom and the sensation avoidance symptom. However, this result must be considered with caution because a study involving a larger sample may highlight other differences.

Third Aspect: The Role of Sensory and Behavioral Symptoms in the RDSP Diagnosis

The third aspect, which is necessary for the validation of the RDSP diagnosis according to the scientific literature, is related to the specific weight of each one of the symptoms in the definition of the RDSP framework [60, 61]. Consequently, it was necessary to analyze how the sensory symptoms contribute to the definition of RDSP. If the role of the sensory symptoms proved to be significant in the RDSP diagnosis and different from the behavioral symptoms, the inclusion of the sensory symptoms in the diagnostic criteria would be justified.

This question incited the authors to analyze the role played by the sensory symptoms together with the behavioral symptoms in the RDSP diagnosis [44]. The hypothesis formulated in this study was that the sensory symptoms were as important as, or even more important than, the behavioral symptoms in the RDSP diagnosis.

To validate this hypothesis, twelve logistic regressions were made, for each of which a sensory and behavioral symptom as predictive variables for RDSP was included. The sensory and behavioral symptoms included in the analyses were those that had been the most discriminative between the groups [41, 44]. These were *Withdrawn behavior* and *Pervasive developmental problems* (PDP) [41, 44]. In line with the methodology of Peduzzi et al. [62], a large number of regressions were necessary because the size of the sample did not yield firm results with more than two variables [62].

The results indicated a significant effect of three sensory functions – hearing, visual, and oral – that increased the probability that the child would fit within an RDSP framework by approximately 3-fold compared to the other diagnoses considered (with associated values of $p \leq 0.008$, as required by Bonferroni [63]). This result occurred in models including PDP and avoiding behaviors.

The weight of the *sensory patterns* did not prove to be as significant in the regression models as the weight of the *sensory areas*. The *sensory patterns* did not approach the level of statistical significance required after applying the Bonferroni correction ($p \leq 0.005$), although in most cases, a clear trend toward significance could be observed ($p \leq 0.05$).

Neither of the two behavioral symptoms analyzed had a significant effect on the RDSP diagnosis. The fact that no significant effect of the behavioral symptoms on the RDSP diagnosis was found is consistent with the theoretical models of sensory integration environments, where the diagnosis is essentially made based on sensory criteria [36,64].

Conclusion

Based on the results of the studies concerning the 3 related aspects regarding the validity of the RDSP diagnosis, the following conclusions that can be drawn:

1. The diagnostic criterion requiring the presence of alterations in sensory modulation to establish an RDSP diagnosis proved discriminative in clinical samples because it was associated with the RDSP diagnosis to a greater extent that with the other diagnoses studied. All of the children with RDSP displayed alterations in sensory functions and patterns, whereas these alterations were only

found in some children with other child psychiatric/psychological diagnoses. This result proves that assessing the presence of alterations in sensory modulation is useful to the RDSP diagnosis.

2. These alterations in sensory modulation were present in a greater number of functions and patterns in children with RDSP and with a greater seriousness of symptoms.
3. The sensory symptoms appeared to be different from the behavioral symptoms, and for this reason, it would be logical to keep both diagnostic criteria as independent aspects.
4. The sensory symptoms are more relevant than the behavioral symptoms in the differential diagnosis of clinical samples in the child psychiatric field included in this study.
5. The results provided by the ITSP [42] contributed to the validation of the sensory diagnostic criterion suggested in the DC:0-3R [10], proved useful in the differential diagnosis of the child psychiatric clinical samples, and converged with the clinical criteria for the RDSP diagnosis.

Clinical Implications

The characteristics that the children with RDSP exhibited in some of the previous studies [41, 44] which were related to severe under-reactivity or over-reactivity to sensory stimuli and the inability to achieve homeostasis (self-regulation), were associated with a less efficient functioning of the autonomic nervous system in a study by McIntosh *et al.* (1999) [36]. This inefficiency of the autonomic nervous system was in turn related to a vulnerability to stress, developmental problems, and emotional and behavioral over-reactivity in children and adults [5, 36, 38].

The differential diagnosis between RDSP and other child psychiatric/psychological diagnoses was fundamentally based on sensory symptoms and only based on behavioral symptoms to a lesser extent. It is thus essential to have instruments that are able to assess such sensory symptomatology in a specific way in mental health and early intervention services that take care of children presenting these characteristics.

Some studies suggest a parallelism between the model of psychiatric/ psychological classification and the model used in occupational therapy.

The practical application of knowledge related to the regulation of sensory stimuli is to create the conditions that facilitate the maturation of structures and affective-emotional regulation in children. This would enable the inclusion of preventive approaches for clinical situations that do not yet comply, given the age of these children, with all of the criteria that are required to address some of the disorders present in the DSM-IV-TR [30]. These disorders, such as Attention Deficit Hyperactivity Disorder [36] or specific emotional or developmental disorders, may be related to these clinical situations, as shown in various studies [4, 6]. The results provided by the ITSP [42] were differential depending on the diagnostic group (RDSP vs. OD3R) and converged with the clinical opinion, thus contributing to the validation of the sensory diagnostic criteria suggested in the DC:0-3R [10].

Future Opportunities for Research

Future studies, conducted with larger samples, would allow for establishing a complete model of logistic regression to study the predictive and discriminative capacity of the group of symptoms characterizing the RDSP diagnosis. A larger sample size would also enable the analysis of the crucial points that may optimize the sensitivity and specificity of the ITSP [42] for the differential diagnosis in clinical samples.

It would be interesting to analyze the stability of RDSP over time. The results of the McIntosh *et al.* study (1999) [36] suggest that it is important to look in older children for the limits of the diagnosis of ADHD according to DSM-IV-TR [30], in particular the type hyperactive/impulsive or combined type. It would also be interesting to analyze the limits among other disorders (i.e.: Anxiety Disorder).

The similarities between Dunn's theoretical model (2002) regarding SMD [42] and the model followed by the DC:0-3R [8] to classify the RDSP subtypes should also be studied to develop a common theoretical framework between occupational therapy and child psychiatry/psychology. It will also be necessary to consider the specifications of the RDSP subtypes that are used in the *Diagnostic Manual for Infancy and Early Childhood: Mental Health, Developmental, Regulatory-Sensory Processing, Language and Learning Disorders* (ICDL-DMIC) [65], which may allow for a common approach between the two classifications to the alterations of sensory processing during early childhood.

In conclusion, the study of sensory processing among children diagnosed with RDSP contributed to the validation of the sensory criterion of one of the most commonly diagnosed frameworks in early childhood [8], which has recently been included in the DC:0-3 [7, 8] and ICDL-DMIC [65] classifications of diagnostic systems. These diagnostic systems consider child psychopathology from a broad point of view, which encompasses child development, the interactions between primary care-givers and infants, and neuroscientific aspects. It is essential to push forward the research on the validation of this framework, which was specifically created for young children, and on the validation of the framework's diagnostic criteria in these classifications [10, 65].

References

[1] Ayres, A. J. (1963). Eleanor Clarke Slagle Lecture – The development of perceptual-motor abilities: A theoretical basis for treatment of dysfunction. *American Journal of Occupational Therapy*, *27*, 221–225.

[2] Miller, L. J., Anzalone, M. E., Lane, S. J., Cermak S. A. & Osten E. T. (2007). Concept evolution in sensory integration: A proposed nosology for diagnosis. *American Journal of Occupational Therapy*, *61*, 135-140.

[3] Posner, M. I. & Rothbart, M. K. (2000). Developing mechanisms of self-regulation. *Development and Psychopathology*, *12*, 427–441.

[4] DeGangi, G. A., Breinbauer, C., Porges, S. W. & Greenspan, S. I. (2000). Prediction of childhood problems at three years in children experiencing disorders of regulation during infancy. *Infant Mental Health Journal*, *21*, 158–175.

[5] DeGangi, G. A., DiPietro, J. A., Greenspan, S. I. & Porges, W. P. (1991). Psychophysiological characteristics of the regulatory disordered infant. *Infant Behavior and Development*, *14*, 37–50.

[6] DeGangi, G. A., Porges, S. W., Sickel, R. Z. & Greenspan, S. I. (1993). Four-year follow-up of a sample of regulatory disordered infants. *Infant Mental Health Journal*, *4*, 330–343.

[7] Zero to Three (1994). Diagnostic classification of mental health and developmental disorders of infancy and early childhood (DC:0-3). Washington, DC: National Centre for Infants, Toddlers, and Families.

[8] American Psychiatric Association (1994). Diagnostic and statistical manual of mental disorders (4th ed.). Washington, DC, EE. UU.: Author.

[9] World Health Organization (1994). International and statistical classification of diseases and related health problems (ICD-10, 10th edn. WHO, Geneva.

[10] Zero to Three (2005). Diagnostic classification of mental health and developmental disorders of infancy and early childhood: Revised Edition (DC:0-3R). Washington, DC: Zero to Three Press.

[11] Skovgaard, A. M., Houmann, T., Christiansen, E., Landorph, S., Jørgensen, T. & Olsen, E. M. et al. (2007). The prevalence of mental health problems in children 1 ½ years of age: The copenhagen child cohort 2000. *Journal of Child Psychology and Psychiatry and Allied Disciplines*, *48*, 62–70.

[12] Wieder, S., Kalmanson, B. & Fenichel, E. (1999). Diagnosing Regulatory Disorders Using DC:0-3: A Framework and a Case Illustration. *Infants and Young Children*, *12*, 79-89.

[13] Dunn, W. (2001). The sensations of everyday life: Empirical, theoretical, and pragmatical considerations, 2001 Eleanor Clarke Slagle lecture. *American Journal of Occupational Therapy*, *55*, 608–620.

[14] Egger, H. & Emde, R. (2011). Developmentally sensitive diagnostic criteria for mental health disorders in early childhood: The Diagnostic and Statistical Manual of Mental Disorders-IV, the Research Diagnostic Criteria-Preschool Age, and the Diagnostic Classification of Mental Health and Developmental Disorders of Infancy and Early Childhood-Revised. *American Psychologist*, *66*, 95-106.

[15] Task Force on Research Diagnostic Criteria: Infancy and Preschool. (2003). Research diagnostic criteria for infants and preschool children: The process and empirical support. *Journal of the American Academy of Child and Adolescent Psychiatry*, *42*, 1504–1512.

[16] Luby, J. L., Heffelfinger, A. K., Mrakotsky, C., Hessler, M. J., Brown, K. M. & Hildebrand, T. (2002). Preschool major depressive disorder: Preliminary validation for developmentally modified DSM-V criteria. *Journal of American Academy of Child and Adolescent Psychiatry*, *41*, 928–937.

[17] Scheeringa, M. S., Zeanah, C. H., Drell, M. J. & Larrieu, J. A. (1995). Two approaches to the diagnosis of posttraumatic stress disorder in infancy and early childhood. *Journal of the American Academy of Child and Adolescent Psychiatry*, *34*,191–200.

[18] Scheeringa, M. S., Zeanah, C. H., Myers, L. & Putnam, F. W. (2002). *Heart rate and RSA reactivity in traumatized preschool children.*

Symposium presented at the International society for traumatic stress studies 18th annual meeting, Baltimore, MD.

[19] Benoit, D., Zeanah, C. H., Boucher, C. & Minde, K. K. (1992). Sleep disorders in early childhood: Association with insecure maternal attachment. *Journal of American Academy of Child and Adolescent Psychiatry*, *31*, 86–93.

[20] Burnham, M. M., Goodlin-Jones, B. L., Gaylor, E. E. & Anders, T. F. (2002). Nighttime sleep-wake patterns and self-soothing from birth to one year of age: A longitudinal intervention study.

[21] Halpern, L. F., Anders, T. F., Garcia-Coll, C. & Hua, J. (1994). Infant temperament: Is there a relation to sleep wake states and maternal nighttime behavior? *Infant Behavior and Development*, *17*, 255–263.

[22] Minde, K., Popiel, K., Leos, N., Falkner, S., Parker, K. & Handley-Derry, M. (1993). The evaluation and treatment of sleep RDCPA 20 disturbances in young children. *Journal of Child Psychology and Psychiatry*, *34*, 521–533.

[23] Sadeh, A. (1994). Assessment of intervention for infant night waking: Parental reports and activity-based home monitoring. *Journal of Consulting Clinical Psychology*, *62*, 63–68.

[24] Chatoor, I., Ganiban, J., Harrison, J. & Hirsch, R. (2001). Observation of feeding in the diagnosis of posttraumatic feeding disorder of infancy. *Journal American Academy of Child and Adolescent Psychiatry*, 40, 595–602.

[25] Chatoor, I., Ganiban, J., Hirsch, R., Borman-Spurrell, E. & Mrazek, D. A. (2000). Maternal characteristics and toddler temperament in infantile anorexia. *Journal of American Academy of Child Adolescent Psychiatry*, 39, 743–751.

[26] Rescorla, L. (2005). Assessment of young children using the Achenbach system of empirically based assessment (ASEBA). *Mental Retardation and Developmental Disabilities Research Reviews*, *11*, 226–237.

[27] Kendell, R. & Jablensky, A. (2003). Distinguishing between the validity and utility of psychiatric diagnoses. *The American Journal of Psychiatry*, *160*, 4–12.

[28] Kendler, K. S. (2006). Reflections on the relationship between psychiatric genetics and psychiatric nosology. *The American Journal of Psychiatry*, *163*, 1138–46.

[29] Robins, E. & Guze, S. B. (1970). Establishment of diagnostic validity in psychiatric illness: its application to schizophrenia. *American Journal of Psychiatry*, *126*, 983-987.

[30] American Psychiatric Association. (2000). Diagnostic and statistical manual of mental disorders (4th ed., text revision). Washington, DC.

[31] Cantwell, D. P. (1975). A model for the investigation of psychiatric disorders of childhood: its application in genetic studies of the hyperkinetic syndrome. En Anthony E. J. (Ed.), Explorations in child psychiatry (57-59). New York: Plenum Press.

[32] American Psychiatric Association. (2013). Diagnostic and statistical manual of mental disorders (5th ed.). Washington, DC.

[33] American Educational Research Association, American Psychiatric Association y National Council on Measurement in Education. (1999). The standards for educational and psychological testing. Washington DC: AERA.

[34] Cesari, A., Maestro, S., Cavallero, C., Chilosi, A., Peccini, P., Pfanner, L. & Muratori, F. (2003). Diagnostic boundaries between regulatory and multisystem developmental disorders: A clinical study. *Infant Mental Health Journal*, *24*, 365–377.

[35] Davies, P. L. & Gavin, J. W. (2007). Validating the diagnosis of sensory processing disorders using EEG technology. *American Journal of Occupational Therapy*, *61*, 176–189.

[36] McIntosh, D. N., Miller, L. J., Shyu, V. & Hagerman, R. J. (1999b). Sensory modulation disruption, electrodermal responses, and functional behaviours. *Developmental Medicine and Child Neurology*, *41*, 608–615.

[37] Schaaf, R. C., Miller, L. J., Seawell, D. & O'Keefe, S. (2003).Children with disturbances in sensory processing: A pilot study examining the role of the parasympathetic nervous system. *American Journal of Occupational Therapy*, *57*, 442–449.

[38] Ben-Sasson, A., Cermak, S. A., Orsmond, A., Carter, A. S & Fogg, L. (2007). Can we differentiate sensory over-responsivity from anxiety symptoms in toddlers? Perspectives of occupational therapists and psychologists. *Infant Mental Health Journal*, *28*, 536-558.

[39] Briggs-Gowan, M. J., Carter, A. S., Bosson-Heenan, J., Guyer, A. E. & Horwitz, S. M. (2006). Are infant-toddler social-emotional and behavioral problems transient? *Journal of the American Academy of Child and Adolescent Psychiatry*, *45*, 849-858.

[40] Cheng, M. & Boggett-Carsjen, J. (2005). Considering sensory processing disorders in the explosive child: Case report and review. *Canadian Child and Adolescent Psychiatry Review*, *14*, 44-48.

[41] Perez Robles, Ballabriga, Diéguez, & Caldeira da Silva. (2012). Validating Regulatory Sensory Processing Disorders Using the Sensory Profile and Child Behavior Checklist (CBCL 1½–5). *Journal of Child and Family Studies*, *21*, 906-916.

[42] Dunn, W. (2002). The infant/toddler sensory profile manual. San Antonio: The Psychological Corporation.

[43] Pérez-Robles, R. (2009). Perfil sensorial para bebés e crianças (7–36 meses). Not published manuscript, Universidad Autónoma de Barcelona, Barcelona.

[44] Pérez-Robles, R., Doval, E., Jané, M. C., Caldeira da Silva, P., Papoila, A. L. & Virella, D. (2013). The Role of Sensory Modulation Deficits and Behavioral Symptoms in a Diagnosis for Early Childhood. *Child Psychiatry and Human Development*, *44*, 400- 411.

[45] Aron, E. N. & Aron, A. (1997). Sensory-processing sensitivity and its relation to introversion and emotionality. *Journal of Personality and Social Psychology*, *73*, 345-368.

[46] Ayres, A. J. (1964). Tactile Functions: Their Relations to Hyperactive and Perceptual Motor Behaviour. *American Journal of Occupation Therapy*, *18*, 221-225.

[47] Ben-Sasson, A., Carter, A. S. & Briggs-Gowan, M. (2010). The development of sensory over-responsivity from infancy to elementary school. *Journal of Abnormal Child Psychology*, *38*, 1193-1202.

[48] Ben-Sasson, A., Carter, A. S. & Briggs-Gowan, M. (2009). Prevalence and correlates of sensory over-responsivity in elementary school. *Journal of Abnormal Child Psychology*, *37*(5), 705-16.

[49] Carter, A. S., Briggs-Gowan, M. J., Jones, S. M. & Little, T. D. (2003). The Infant-Toddler Social and Emotional Assessment (ITSEA): Factor structure, reliability, and validity. *Journal of Abnormal Child Psychology*, *31*, 495–514.

[50] Dunn, W. (1997). The impact of sensory processing abilities on the daily lives of young children and their families: A conceptual model. *Infants and Young Children*, *9*, 23–35.

[51] Goldsmith, H. H., Van Hulle, C. A., Arnerson, C. L., Schreiber, J. E. & Gernsbacher, M. A. (2006). A population-based twin study of parentally reported tactile and auditory defensiveness in young children. *Journal of Abnormal Child Psychology*, *34*, 393–407.

[52] Kagan, J. & Snidman, N. (1991). Infant predictors of inhibited and uninhibited profiles. *Psychological Science*, *1*, 40-44.

[53] Kinnealey, M. & Fuiek, M. (1999). The Relationship between Sensory Defensiveness, Anxiety, Depression and Perception of Pain in Adults. *Occupational Therapy International*, *6*, 195-206.
[54] Miller, L. J., Coll, J. R. & Schoen, S. A. (2007). A randomized controlled pilot study of the effectiveness of occupational therapy for children with sensory modulation disorder. *American Journal of Occupational Therapy*, *61*, 228–238.
[55] Ben-Sasson, A., Carter, A. S. & Briggs-Gowan, M. J. (2009). Sensory over-responsivity in elementary school: Prevalence and social-emotional correlates. *Journal of Abnormal Child Psychology*, *37*, 705–716.
[56] Gouze, K. R., Hopkins, J., LeBailly, S. A. & Lavigne, J. V. (2009). Re-examining the epidemiology of sensory regulation dysfunction and comorbid psychopathology. *Journal of Abnormal Child Psychology*, *37*, 1077–1087.
[57] Gunn, T. E., Tavegia, B. D., Houskamp, B. M., McDonald, L. B., Bustrum, J. M. & Welsh, R. K. (2009). Relationship between sensory deficits and externalizing behaviors in an urban, Latino preschool population. *Journal of Child and Family Studies*, *18*, 653–661.
[58] Van Hulle, C. A., Schmidt, N. L. & y Goldsmith, H. H. (2012). Is sensory over-responsivity distinguishable from childhood behavior problems? A phenotypic and genetic analysis. *The Journal of Child Psychology and Psychiatry*, *53*, 64-72.
[59] Achenbach, T. M. & Rescorla, L. (2000). Manual for the ASEBA preschool forms & profiles. Burlington, VT: University of Vermont Department of Psychiatry.
[60] Bar-Shalita, T., Vatine, J. J., Seltzer, Z. & Parush, S. (2009). Psychophysical correlates in children with sensory modulation disorder (SMD). *Physiology y Behavior*, *98*, 631-9.
[61] James, K., Miller, L. J., Schaaf, R., Nielsen, D. M. & Schoen, S. A. (2011). Phenotypes within sensory modulation dysfunction. *Comprehensive Psychiatry*, *52*, 715–724.
[62] Peduzzi, P., Concato, J., Kemper, E., Holford, T. R. & Feinstein, A. R. (1996). A simulation study of the number of events per variable in logistic regression analysis. *Journal of Clinical Epidemiology*, *49*, 1373-1379.
[63] Bender, R. & Lange, S. (2001). Adjusting for multiple testing—when and how? *Journal of Clinical Epidemioly*, *54*, 343–349.
[64] Williamsom, G. G. & Anzalone, M. E. (2001). Sensory integration and self-regulation in infants and toddlers: Helping very young children

interact with their environment. Washington, DC: Zero to three. National Center for Infants Toddlers and their families.

[65] Interdisciplinary Council on Developmental and Learning Disorders (ICDL). (2005). Diagnostic manual for infancy and early childhood. Bethesda, MD: Interdisciplinary Council on Developmental and Learning Disorders.

Index

#

A

B

C

F

G

H

I

K

L

M

Q

R

S

T

U

V

W

Y